GEOLOGICAL HAZARDS

NIPA® GENX ELECTRONIC RESOURCES & SOLUTIONS P. LTD.
New Delhi-110 034

Geological Hazards

Causes, Consequences & Methods of Containment

Edited by

Dr. Mu. RAMKUMAR

Department of Geology
Periyar University, Salem - 636 011
Tamil Nadu, India

NIPA® GENX ELECTRONIC RESOURCES & SOLUTIONS P. LTD.
New Delhi-110 034

NIPA® GENX ELECTRONIC
RESOURCES & SOLUTIONS P. LTD.

101,103, Vikas Surya Plaza, CU Block
L.S.C. Market, Pitam Pura, New Delhi-110 034
Ph : +91 11 27341616, 27341717, 27341718
E-mail: newindiapublishingagency@gmail.com
www: www.nipabooks.com

For customer assistance, please contact
Phone: + 91-11-27 34 17 17
Fax: + 91-11- 27 34 16 16
E-Mail: feedbacks@nipabooks.com

ISBN: 978-81-19215-48-5

Composed and Designed by NIPA®.

Foreword

Our planet, the earth is unique as it supports life and allows it to evolve and flourish. Life thrives on this planet despite geohazards, which are inalienable part of planetary processes through the ages. Geohazard can be defined as a geological state that represents or has the potential to develop further into a situation leading to damage or uncontrolled risk. Events of destruction and reconstruction have been a part to the earth's history and affect even the normal activities of life since the advent of industrial era and more so in the current era of globalisation among our mechanized-networked-automated-populated society.

Prior to the development of scientific thought, civilizations all over the world ascribed geohazards as acts of God. Mythologies world over are full of stories connecting various geological events to supernatural power. The Hindu mythologies consider the earth as the mother and each morning before stepping upon the earth, a devout Hindu chants the following mantra: Wearing attire of the oceans, goddness with mountain like breasts; wife of Lord Vishnu, OH the mother earth, I bow and touch thy feet, please pardon me. Human reasoning and logic took thousands of years to realize that geohazards are caused by the nature's process. With the population explosion and globalization, lifestyles have changed from "in tune with nature" to "tailored to fit the needs of economy" which in turn pose constraints on natural resources, affecting the natural geological processes, creating hazardous situations. Nowadays geohazards have become widespread phenomena that are related to geological and environmental conditions and involve long-term and/or short-term geological processes. It demonstrates the effects of extreme geological phenomena operating as part of the earth systems on humans. As population growth forces more communities to expand into areas at risk from these ominous threats escpecially in the third world countries, concern increases about the danger that geohazards pose to people, property, and the environment.

Geohazards can thus be relatively small features, but they can also attain huge dimensions and affect local and regional socio-economy to a large extent. In addtition, human activities could also result in significant risk, and as such mitigation and prevention are paramount, through improved understanding of geohazards, their preconditions, causes and implications. The risk is increasing with the human population and the growth of the technical world, the high sophisticated equipments, needed for our daily life such as infrastructure (e.g. roads, railroads, pipelines, electrical transmission, and telecommunication cables, etc.). Every year geohazards such as volcanoes, earthquakes, landslides and tsunamis claim thousands of lives devastate homes and destroy livelihoods.

Geohazards can be of different scales in dimension and time and of different complexity in structure. Since several years many scientists have therefore focused their research on geohazards and have developed powerful new approaches to addressing the problems of geohazards risk assessment and mitigation. Often multi-disciplinary investigations are needed for better observation, for long-term awareness of geohazards, visualization, for interpretation and understanding relevant processes, planning, monitoring through satellite imagery and onsite measurements and for establishing specific mitigation studies and relevant prevention mechanisms such as disaster scenarios, forecasts, and post-event assessment maps.

Geohazards are part of the earth's environment and they cannot be wished away. Only way to tackle them is to know the answers about why do we encounter such hazards and when and how they occur. This knowledge can help the society to remain prepared for the unexpected and escape their impact. This book therefore is a very good approach, even if there are several more books coming out soon on various aspects of this topic. Presentation of comprehensive information on common types of geohazards and detailing different types of geohazards in the context of dynamics of population and geoenvironment are appreciable and may enrich the knowledge of the readers.

Prof. Dr.Doris Stueben
Institute of Mineralogy and Geochemistry
University of Karlsruhe, Germany

Message

Geohazards are natural geological processes that affect the human community. Some times they become disastrous due to the interference by the humans. However, these hazards can be mitigated with proper planning. Failure in implementing a scientific mitigation strategy will result in a situation that happened on 26th December 2004 along the eastern coast of India. The 26th December tsunami has brought in awareness among the administrative and other governmental agencies to draw a scientific programme with the help of the world scientific community to demarcate zones vulnerable for such disasters, assess risks and prepare for future such eventualities. Proper planning will help in mobilizing rescue and relief facilities to the affected.

With the open economy and the need to keep pace with economic development of other countries, utilization of natural resources to the benefit of mankind and preservation of the natural environment without any negative effects are mandatory objectives. In developing economies such as India, values attached to resources are more than the human lives that results in habitation of vulnerable regions without any thought on safety of lives and infrastructure. This is how huge fatalities and economic losses result in developing countries than the developed countries for similar type and intensity of geohazards. Although India is endowed with abundant land, water, mineral and biological resources, the very nature of uneven distribution these resources has created a situation that ecologically sensitive, environmentally fragile and geologically dynamic regions are extensively put to use without any consideration towards conservation of natural equilibrium and safeguarding the interests of human endeavors, which, often results in calamities.

Past experiences suggests that creation of better awareness on natural geological process and causative factors for disasters are important in averting loss to live and property. Mitigation strategies certainly will save millions from such disasters. In order to achieve this, it is important to map hazardous zones, assess their impact on the environment and lives and property, draw hazard preparedness and implement safety norms. Among these keys factors, creation of better awareness is of paramount importance and is a first step that may lead to the creation of a safe and prosperous country. Realizing this importance, governmental and non-governmental agencies are taking initiative to popularize nature and characteristics of geohazards and strategies to minimize calamities. Inclusion of information on geohazards in school syllabus, conducting periodic awareness and training programmes and mass media campaign are some examples of this initiative. Despite these efforts, it should be admitted that still a long way has to go in accomplishing these objective. Many reports, research publications and special volumes are published on geohazards, but a single resource book providing all the fundamental, yet accurate and easy to comprehend information is still lacking. This goal is achieved by the present book that contains information on several types of geohazards.

Nineteen chapters, offering information on various geohazards, causative factors, and policy initiatives required to contain ill effects of geohazards, management strategies that need to be implemented for mitigation of geohazards and a specific section on glossary of terminologies related to geohazards are presented in this book. Written in a simple, yet highly informative style, this book should serve as a single point academic resource pertaining to geohazards. It is a complement to the authors for having brought such a book at a time when the world is facing several natural disasters. This book can serve as a text book for the students of geosciences and a guide to all those concerned with safety of natural environment, planners and administrators and common public.

Prof. D.Chandrasekharam
Department of Earth Sciences
IIT – Bombay, Mumbai - 400076

Preface

Recurrence of geological hazards such as cyclones, floods, tsunami, earthquake and landslides and resultant life and property loss in the recent past has instilled fear among common public. Although the governmental and other agencies work for the betterment of quality of life and protection of natural environment, lack of clear understanding on natural geological hazards and their nature of impact create confusion leading to mismanagement of relief and rescue operations. In this regard, it is felt that making available a single source of information pertaining to types, causes, consequences and mitigation of geohazards would be an ideal effort to create better awareness among common public, planners and administrators.

According to a report by the Government of India, lack of capacity to limit the impact of hazards remains a major burden for developing countries. An estimated 97% of natural disaster related deaths each year occur in developing countries and, although smaller in absolute figures, the percentage of economic loss in relation to the Gross National Product (GNP) in developing countries far exceeds the ones in developed countries. While only the direct financial impacts of a disaster such as damage to infrastructure and agricultural produce could be estimated, the hidden costs in terms of loss of personal belongings or jobs, widening trade or government budget deficits, or the increasing poverty, are even higher.

Hazard is described as a cause due to which disasters happen, vulnerabilities of which vary according to the type and intensity in addition to various other reasons such as topography, slope, soil and lithology etc. Vulnerability is defined as "the extent to which a community, structure, service, or geographic area is likely to be damaged or disrupted by the impact of particular hazard, on account of its nature, construction and proximity to hazardous terrain or a disaster prone area." Risk is a measure of potential impact in the event of such a calamity. Disaster Preparedness is the level of anticipation and fortification against geohazard. Thus, hazard, disaster, vulnerability, risk and preparedness are the five terms that need to be understood and hence individual chapters in this book deal with specific geohazard, cause, potential impact and mitigation measures for the understanding of general reader. Although many sources could supply such information, presentation of facts pertaining to individual disasters in a simple, yet comprehensive way enhances the potential of this book to create better awareness on geohazards. As explained in many of the chapters of this book, better awareness is the pre-requisite to save lives and property from the impacts of geohazards. Mitigation measures for geohazards are dependent on local conditions and there are plenty of published literatures in this regard. For this reason, each chapter has devoted a section on mitigation measures presenting only the policy initiatives and broad spectrum of activities required for containing adverse impacts. Many chapters point out the directives to be undertaken in India by the planners and administrators besides the common public.

Dr. Mu. RAMKUMAR

Acknowledgements

This book contains a collection of chapters authored by many experts who have toiled hard to find causes and solutions to natural hazards and those who had shared the feeling of the editor that increased awareness among common public, administrators and planners on type and nature of natural hazards would save many a lives, property and environment. In this context, I thank all those who have contributed towards presentation of concise, yet all the critical information on natural geological hazards *a la* geohazards.

Former Vice-Chancellor of Periyar University, Prof.Dr.T.Balakrishnan, is thanked for his administrative and moral support during whose tenure the idea of conducting research on geohazards and mapping vulnerable areas by the editor of this book was conceptualized and initiated. Prof.Dr.T.Balakrishnan was the person who has made the editor to see reason in disseminating information on geohazards through mainstream media for the benefit of common public. This book is an effort to create increased awareness among common public, the seedlings of which were planted during his tenure. Honorable Vice-Chancellor, Prof.Dr.M.Thangaraju, Prof.Dr.S.Gunasekaran, Registrar and staff of Department of Geology, Periyar University, are thanked for their support. Typographic and other helps extended by G.Sathish, T.Sugantha, S.Karthikeyan and M.Karthikeyan, Research Scholars, Department of Geology, Periyar University are gratefully acknowledged.

Many individuals, either directly or indirectly and either in their official capacity or as a goodwill gesture helped the editor in compiling this book. This acknowledgement section would not be complete without thanking my wife Mrs.A.Shanthy, daughter Ms.Ra.Krushnakeerthana and son Mr.Ra.Shreelakshminarasimhan, who have helped passively by allowing me to edit peacefully, by foregoing my presence in their hours of need and offering their comments on contents of few chapters as general readers.

I thank the reviewers, who chose to remain anonymous, for having helped the editor in the task of separating chaff from grains without altering the style and content of the chapters. New India Publishing Agency, New Delhi, is thanked for printing this book on time with eloquent style and design.

Dr. Mu. RAMKUMAR

Contents

1

Types, Causes and Strategies for Mitigation of Geological Hazards

Mu.RAMKUMAR
Department of Geology, Periyar University, Salem - 636 011.
E-Mail : <muramkumar@yahoo.co.in>

In this chapter, an attempt is made to emphasize the fact that while no country in the World is safe from geohazards, a proper understanding on their causes, ways to mitigate them and surviving techniques would help reduce the life and property loss. Each geohazard has its own characteristic way of potential impact and hence classifying vulnerability of different regions and creating awareness on potential impact is essential. In addition to providing basic information on geohazards, this chapter presents a conceptual model on hazard mitigation, rescue and relief operations.

INTRODUCTION

Every year, thousands of people are killed by volcanic eruptions, earthquakes, landslides and floods. These natural causes damage human settlements and infrastructures (GOI, 2004) inflicting life and economic loss. As human population increases, habitation on hazardous and or vulnerable land becomes more common and the risks posed by these hazards increase. The need to observe their behavior, understand them better and mitigate their effects become ever more urgent (IGOS, 2003). Decades of research may have contributed towards betterment of our understanding on the processes of geohazard formation but still a long way has to be covered to develop better mitigation procedures.

Wikipedia (www.wikipedia.com) describes geologic hazard as an adverse geologic condition that happens suddenly or slowly and capable of causing damage or loss of life and property. Arya et al. (2005) define hazard as a cause due to which disasters happen, vulnerabilities of which vary according to the type and intensity. The vulnerability of an area may vary depending on many reasons including proximity, topography, slope, structure, vegetation, etc. Vulnerability is defined as "the extent to which a community, structure, service, or geographic area is likely to be damaged or disrupted by the impact of particular hazard, on account of its nature, construction and proximity to hazardous area or a disaster prone area." Risk is a measure of potential loss in the event of calamity at a particular region for a certain period of time. Disaster preparedness is the level of fortification in terms of relief and rescue operations. Preparedness includes all the measures taken before a disaster event, aimed at minimizing loss.

Geological hazards or natural hazards, termed collectively as geohazards are those dangerous geological events that occur due to natural processes without any external catalyst or as a result of anthropogenic intervention with natural geological equilibrium/ processes of an area/region, the potential impacts of which may be felt upto few meters to many hundreds or thousands of kilometers from the point of origin of catastrophe. Examples in this regard are: land subsidence (localized caving in of land running for few meters resulting in collapse of few houses) and the recent tsunami (that destructed the whole of coastal regions of south Asia) or volcanic explosion in Mount St.Helens (death of thousands of people and complete destruction of infrastructure in the vicinity). In view of their severity, unpredictability and potential to wreck havoc over large areas, these geological events have to be understood clearly.

According to a report by the Government of India-UNDP Disaster Risk Management Program (Arya et al. 2005) while no country in the world is entirely safe from geohazards, lack of capacity to limit the impact of hazards remains a major burden for developing countries including India. An estimated 97% of natural disaster related deaths each year occur in developing countries and, although smaller in absolute figures, the percentage of economic loss in relation to the Gross National Product in developing countries far exceeds the ones in developed countries. Natural disasters are potentially serious shocks to an economy. While direct financial impacts of a disaster such as damage to infrastructure and agricultural produce could only be estimated, the hidden costs in terms of loss of personal belongings or jobs, widening trade or government budget deficits, or the increasing poverty, are even higher. The urgency to improve awareness levels on geohazards stems from the fact that 70% of the cultivable lands of India is prone to droughts, 60 % of land area is prone to earthquake hazards, 12% of area is prone to flooding and 8% of area is subjected to cyclones annually.

Under this scenario, it is imperative that people - common public, administrators and planners alike, understand the types and nature of important geohazards, modalities to mitigate them and measures to be undertaken in the event of emergency for conducting relief and rescue operations - at least to minimize loss of life and damage to property. As impending catastrophes could not be predicted and the nature of geohazards are such that an earthquake happened at thousands of kilometers away could wreck havoc in such a large region as that of southeast Asia in the form of recent tsunami, increased awareness is the need of hour that may save precious lives directly and help improve economy of our country by minimizing costs of damage control. This chapter is an attempt to introduce the reader to various types of geohazards, and present a workable model of mitigation, forecasting, relief and rescue operations.

GEOLOGICAL HAZARDS

Earthquake

Earthquakes are one among the highly dangerous geological hazards. They pose severe threat due to their unpredictability and associated secondary impacts. Earthquakes are violent vibrations of the earth plate, caused by sudden release of energy that build-up at and/or beneath the surface of the earth crust. It is of common knowledge that the earth is crust is made up of many plates floating over colloidal-viscous molten magma called mantle. The plates are at places collisional (convergent margin) and in other places move away (divergent margin) from each other. Whether convergent or divergent, these plate boundaries, associated faultlines and related structures create seismicity, the intensity of which is measured in terms of the magnitude of seismic waves. There are many types of seismic waves that travel along the surface and across the surface, all of which are detrimental to man-made structures. The point of origin of earthquake at surface is called "epicenter". The deep-seated point located directly below the epicenter is called "focus" from where, the seismic waves propagate radially in all directions. The intensity of earthquake is measured in terms of Richter or Mercalli scale. India Meteorological Organization classifies the earthquakes according to the magnitude viz., upto 4.9 in Richter scale - "slight"; 5-6.9 as "moderate"; 7-7.9 as "great" and 8 and above as "very great". It is to be noted that magnitude of 2 is ten times greater than that of magnitude of 1. Hence, the destructive force of the earthquake from one level to other level would be ten times than that of next lower scale.

Earthquakes generate allied geohazards namely landslides, tsunami, etc. Although a natural phenomenon, earthquakes can also happen due to volcanic explosion, landslides, nuclear explosion, mine blasts, etc. During the past 25 years, earthquakes have caused more than 1 million deaths worldwide. Seventy percent of the earthquakes measuring seven or over on the Richter scale occurred in the Asian and Pacific region, at an average rate of 15 per year. The most devastating earthquake in the World in recent history, the Tangshan earthquake, which occurred in China on 28th July 1976, is reported to have claimed over 2,40,000 lives. The region in and around Japan, Indonasia, Java and Sumadra Island is one such seismicity prone area that constantly experiences minor to major earthquakes. While the Japanese islands and the surrounding continental shelves cover only 0.1 per cent of the total area of the world, a whopping 10% of total earthquake energy emanates from this area alone. In Japan, a great earthquake of Richter scale 8 recurs every ten years, and a large-scale earthquake of magnitude 7 once a year.

About 65% of the total area of India is vulnerable to seismic damage of buildings in varying degrees. India had witnessed several earthquakes like the Uttarkashi earthquake of 1991, the Latur earthquake of 1993, the Jabalpur earthquake of 1997 and the Chamoli earthquake of 1999. The Bhuj earthquake of 26th January 2001 and the Jammu and Kashmir earthquakes of 8th October 2005 followed these. One notable feature of these earthquakes is that the casualties were caused primarily due to the collapse of buildings. However, similar earthquakes in the United States, Japan, etc., do not lead to such enormous loss of lives, as the structures in these countries are built with structural mitigation measures and earthquake-resistant features. This emphasizes the need for strict compliance of town planning byelaws and earthquake-resistant building codes in India (Arya *et al.* 2005).

Landslide

Landslides rank next to earthquakes in causing fatalities and economic/ infrastructure loss. A wide variety of ground movements principally driven by gravity along a sloping terrain, located anywhere from mountains to coastal and deep sea can be termed as landslide. Nevertheless, other factors such as slope instability created by earthquakes, volcanic explosion, blasting, loss/sudden removal of vegetal cover on a sloping terrain, variation of hydrostatic and pore water pressure, erosion by rivers, glaciers and waves, heavy rain, liquefaction, etc, can aid in and/or initiate landslide. Wikipedia suggests that pre-conditional factors build up specific sub-surface conditions that make the area/slope prone to failure, actual landslide often requires a trigger. While movement of large mass of earth material is termed landslide, sudden movement of glacial ice and rock material downslope is termed as "avalanche" and downslope movement of hot ash, gas and rocks from a volcanic explosion is termed as "pyroclastic flow". Submarine slope failure and movement of sediment pile towards deeper regions is termed as "mass wasting".

Landslide often takes place in conjunction with or triggered by earthquakes, flooding, heavy rainfall and volcanic activity. Bhandari (2006), in his classic review on landslides stated that normally landslides move at the rate of 1-4 cm per day. However, high-speed landslides including mudflows, debris flows and debris avalanches, associated with heavy rainfall or earthquakes, or both move at speeds as high as 3 to 50 m/s. According to him, volume of the land mass displaced, suddenness, recurrence and momentum dynamics are the factors deciding on the degree of severity of a landslide.

Although occurrences of landslides during geological past are known, periodicities and number of incidences of landslides have increased many folds since the advent of industrial era and rapid urbanization, principally due to unscrupulous alteration of natural slopes, geomorphic conditions and land-use pattern. As all these anthropogenic activities result in landscape modification, removal of vegetal cover, increase in slope instability and increase in pore water pressure, incidences of landslides is on the rise. In India, the Himalayan and Western Ghat regions are considered to be most vulnerable to landslides. Between these two, occurrences of landslides are frequent in Himalayan region.

Tsunami

Tsunami is a word (spelt as "Soo-NAH-Mee") originated from Japanese language meaning "harbor wave." *Tsu* means harbor and *nami* stands for wave. Tsunamis are large waves generated during violent shaking of sea floor and vertical displacement of overlying watercolumn by seismic activity in the ocean. Tsunamis frequent coastlines located adjacent to major submarine earthquake prone zones. The Hawaii islands and Japanese island regions are the active locales of tsunami formation. When submarine earthquake shakes the overlying seawater, the seawater above the epicenter raises to a few centimeters high which moves towards all the directions. During its journey, height of the wave rises, which assumes enormous proportions when it reaches continental shelves and attains heights upto 30 meters at breaker zone. The travel speed of the tsunami varies widely according to the magnitude of earthquake, submarine bottom topography, etc., from few meters to few hundreds of kilometers per hour. The speed of the wave current and its incremental rate while breaking at coastline are the two main factors that decide the severity of destruction. Although submarine earthquake is considered to be the causative factor of tsunami generation, other factors such as submarine volcanic explosion, mass wasting and even meteoritic impact could generate tsunamis. In geological history, the buried Chixlub impact crater near Mexico created 65 million years ago is considered to be the mother of all tsunamis known till today.

Tsunamis are very common over the Pacific Ocean because it is surrounded on all sides by a seismically active belt. In the Hawain Islands, Tsunamis approach from all directions, namely, from Japan, the Aleutian Islands and from South America (Arya et al. 2005). The Pacific-wide tsunami generated by Chilean earthquake of 22nd May 1960 was considered to be the most destructive tsunami prior to the Tsunami of 26th December 2004 in Indian Ocean region. The Indian coastal belt has not recorded many Tsunamis in the past. Waves accompanying earthquake activity have been reported over the North Bay of Bengal. During an earthquake in 1881, which had its epicenter near the centre of the Bay of Bengal, Tsunamis were reported. The earthquake of 1941 in Bay of Bengal caused some damage in Andaman region. During the earthquakes of 1819 and 1845 near the Rann of Kutch, there were uncommon large waves.

Volcanic Eruption

Volcanoes are geomorphological features located along active plate margins either in continental and submarine regions spewing hot molten magma, toxic gas and pyroclastic materials. Volcanoes are perforations in the earth's crust through which molten rock and gases escape to the surface and atmosphere. Volcanic hazards stem from two classes of eruptions namely, "explosive" and "effusive". Explosive eruptions originate in the rapid expansion of gas from the molten rock as it nears the earth's surface. Explosions pose a risk by scattering of hot rock blocks and fragments and lava at varying distances from the source. The volcanic materials flow instead of explosion in effusive type of volcanism. Flows vary in nature (mud, ash, lava) and quantity and may originate from multiple sources. Flows are governed by gravity, surrounding topography and viscosity of ejecta.

Although locations of volcanoes are precisely mapped, unpredictability of eruptions makes volcanoes one of the most devastative geohazards. A variety of eruptions from volcanic mountains are observed that range from gentle pouring of lava from volcanic vent to violent explosive eruptions. Air borne ash and pyroclastic materials, avalanche of hot gas and molten magma, volcanic mud flows (volcanic ash mixed with water) are introduced during eruptions that pose severe and sudden threat to the areas adjacent to volcanoes. The area affected by eruptions range from few hundreds of meters to few thousands of kilometers. Sudden eruptions of dormant volcanoes that were covered by thick pile of glacial ice pose additional threat of flooding. Volcanoes are classified according to their eruption intervals and also the type of eruption. Violent volcanic eruption causes earthquakes and landslides and also large-scale avalanche in continental and submarine regions. Submarine volcanoes are recorded to have generated tsunamis too.

Volcanoes are known to have caused collateral damage since historic times. Volcanic eruption in western India during 65 million years ago is considered to be one of the greatest in earth's history that might have caused or contributed towards extinction of many fauna and flora either directly or indirectly by creating environmental stress.

Flood

United States Geological Survey defines flood, as "relatively high stream flow that overflows the natural or artificial banks of a stream or that submerges land not normally below water level. Arya et al. (2005) define floods as a temporary inundation of large regions as the result of an increase in reservoir, or of rivers flooding their banks because of heavy rains, high winds, cyclones, storm surge along coast, tsunami, melting snow or dam bursts. Two types of flooding can be distinguished namely,

- Land-borne floods, or river flooding, caused by excessive run-off brought on by heavy rains, and
- Sea-borne floods, or coastal flooding, caused by storm surges, often exaggerated by storm run-off from the upper watershed.

Flooding is a recurrent disaster that affects almost every part of India. Flooding impact is particularly high in coastal and deltaic regions owing to hapazard urbanization, reduction of carrying capacity of streams due to heavy siltation and inadequate drainage network to transport meteoric water to streams. During flooding, supply routes and other infrastructure get affected due to which flooding in India also produces endemic diseases that aggravates the severity of flood impact, particularly health hazards to fauna and flora including humans. Almost all the Indian rivers receive heavy influx during June to September and this period is also the spell of flooding disaster in India. Although Indian sub-continent has diverse topographic features ranging from glacial mountains to coastal plains, flooding in India is ubiquitous namely, the northern mountainous regions experience flooding due to snow melt floods, the coastal regions get flooded due to storm surges associated with cyclones and thunderstorms. Central water commission, Ministry of water resources, Government of India has estimated that annually an amount of Rs.2706.243 crore is lost by flooding impact.

Flood disaster management practices in India is targeted towards improving carrying capacity of the rivers, desiltation measures, construction of structural walls along river banks and improving drainage network systems, diversion canals, etc. Flood forecasting, contingency plans for rescue and relief measures are also being implemented.

Drought

In a concept paper of World Meteorological Organization (WMO), Jarraud (2006) describes that throughout the course of human history, drought has been a problem affecting human welfare and food security. Drought is a temporary reduction in water or moisture availability significantly below the normal or expected amount for a specific period. It results either due to inadequacy of rainfall, or lack or irrigation facilities, over-exploitation or deficient availability for meeting the normal crop requirements in the context of the agro-climatic conditions prevailing in any particular area. Drought is a slow-onset disaster and it is difficult to demarcate the time of its onset and end. Falling rainfall and groundwater levels, drying wells, rivers and reservoirs and poor agricultural production indicate onset of drought. According to the India Meteorological Department, 10% or more deficiency of rainfall of a region is considered to be drought condition in that region. Drought by itself is not a disaster. Its transformation into a disaster depends on its impact on local people, economies and the environment and their ability to cope with and recover from it. Jarraud (2006) classifies droughts into four types namely, meteorological, agricultural, hydrological and socioeconomic. The interrelationship between these types of droughts has three distinguishing features namely, intensity, duration and spatial coverage. Droughts can develop quickly in some climatic regimes, but usually require a minimum of two to three months to become established. Once begun, it can persist for months or years.

The Indian subcontinent experiences an average rainfall of around 1200 mm and as such its water resources are enormous and are well comparable to any other country of its size. However, the problem lies in its distribution across the country, ranging from over 10,000 mm at cherrapunji and 8500 mm in parts of Western Ghats at one end to around 200-350 mm in parts of western Rajasthan and Gujarat at the other. Because of the erratic behavior of the rainfall, even the medium rainfall region is vulnerable to drought conditions. Consequently, 68% or roughly 2/3 of the country's arable area is at one time or the other susceptible to drought. Most of the major states have both drought as well as high rainfall areas and there are situations in states like Bihar, where while the northern part reel under floods, some of the southern districts falling in the rain shadow simultaneously experience drought. In addition to geographic distribution factor of rainfall, the seasonality of rainfall is also favoring drought conditions in India. Most of the precipitation occurs within a span of four months and rest of the period of a year happens to be dry months. Inadequate facilities to store the rainfall excess lead to drought conditions in many parts of the country.

Cyclone

The word "Cyclone" is derived from the Greek word "Cyclos" meaning the coils of a snake. United States Geological Survey defines hurricanes as "Severe cyclones, or revolving storms, originating over the equatorial regions of the Earth, accompanied by torrential rain, lightning, and winds with greater speed. As cyclones bring in gusty winds, which influence sea level to swell and cause coastal inundation, storm surges and flooding in inland rivers, the impacts are multifold. Cyclones are intense low-pressure areas - from the centre of which pressure increases outwards. The amount of the pressure drops in the center and the rate at which it increases outwards gives the intensity of the cyclones and the strength of winds. A full-grown cyclone is a violent whirl in the atmosphere 150 to 1000 km across, 10 to 15 km high. Gale winds of 150 to 250 kmph or more spiral around the center of very low-pressure area with 30 to 100 hPa below the normal sea level pressure. The central calm region of the storm is called the "Eye". The diameter of the eye varies between 30 and 50 km and is a region free of clouds and has light winds. Around this calm and clear eye, there is the "Wall Cloud Region" of the storm about 50 km in extent, where the gale winds, thick clouds with torrential rain, thunder and lightning prevail. Table 1 shows the classification of cyclones by World Meteorological Organisation (WMO) based on the strength of winds.

Table 1. Cyclone classification based on Wind speed

Types of Disturbances	Associated wind speed in the Circulation
Low Pressure Area	Less than 17 knots (< 31 km/hr.)
Depression	17 to 27 knots (31 to 49 km/hr.)
Deep Depression	28 to 33 knots (50 to 61 km/hr.)
Cyclonic Storm	34 to 47 knots (62 to 88 km/hr.)
Severe Cyclonic Storm	48 to 63 knots (89 to 118 km/hr.)
Very Severe Cyclonic Storm	64 to 119 knots (119 to 221 km/hr.)
Super Cyclonic Storm	120 knots and above (222 km/hr. and above)

India has a very long coastline of 8041 km, major portion of which is vulnerable to cyclone. Cyclones in India strike the east and west coast although the frequencies of cyclones along east coast outnumber the westcoast. Out of the storms that develop in the Bay of Bengal, over 58 percent approach or cross the east coast in October and November. According to the India Meteorological Organization (IMD), the following segments of eastern India are most vulnerable to cyclones and associated high surges: North Orissa, and West Bengal coasts, Andhra Pradesh coast between Ongole and Machilipatnam, Tamil Nadu coast, Maharashtra coast, south Gujarat coast and the coastal belt around Bombay and the Gulf of Kutch.

Cyclones are associated with high-pressure gradients and consequent strong winds. These, in turn, generate storm surges. A storm surge is an abnormal rise of sea level near the coast caused by a severe tropical cyclone; as a result, seawater inundates low-lying areas of coastal regions to cause destruction. Very strong winds may damage installations, dwellings, communication systems, trees, etc., resulting in loss of life and property. Heavy and prolonged rains due to cyclones may cause river floods and submergence of low-lying areas by rain causing loss of life and property. Floods and coastal inundation due to storm surges pollute drinking water sources causing outbreak of epidemics.

IMD monitors the atmospheric and other disturbances and sends weather bulletins and cyclonic warnings periodically to public and official machinery to take precautionary measures.

Forest Fire

Although forest fire as a natural hazard is defined as "combustion, marked by flames or intense heat, in natural settings, often ignited by lightning" almost 90 % of forest fires are ignited by human activities (Arya et al. 2005). Forest fires are of two types namely, "surface fire" and "crown fire". A forest fire may burn as a surface fire, spreading along the ground consuming surface and/or burning crown of trees and shrubs, often sustained by a surface fire. A crown fire is particularly very dangerous in a coniferous forest because resinous material given off burning logs burn furiously. On hill slopes, if the fire starts downhill, it spreads up fast as heated air adjacent to a slope tends to flow up the slope spreading flames along with it. If the fire starts uphill, there is less likelihood of it spreading downwards. Forest fires pose a threat not only to the forest wealth but also to the entire regime to fauna and flora seriously disturbing the bio-diversity and the ecology and environment of a region. During forest fires, natural vegetation, wildlife habitat, natural biodiversity are lost in addition to negative impacts such as acceleration of soil erosion, reduction in atmospheric moisture, increase in atmospheric CO_2 and contribution towards global warming.

In their vivid review, Arya *et al.* (2005) stated that as per the statistics of Forest Survey of India, the actual forest cover of India is 19.27% of the geographic area (accounting for 1% of World's forest resources) corresponding to 63.3 million ha, in which only 38 million ha of forests are well stocked with a crown density of about 40%. This resource has to meet the demand of a population of 950 million people and around 450 million cattle. As such, our country has to meet the needs of 16% of the world's human population and 19% of World cattle population from 1% of the world forest resources. The forests of the country are therefore, under tremendous pressure. Forest fires are a major cause of degradation of India's forests. The normal fire season in India is from the month of February to mid June. India witnessed the most severe forest fires during the recent times, causing a loss of forest area of 677,700 ha.

Meteorite Impact

Nevertheless discounted as a minor and remotely possible event, consideration of meteoritic impacts as natural hazard has emerged important since the landmark article of Grieve (1990), which has pointed out the magnitude of impacts that may arise out of meteoritic impact. Grieve's (1990) views are presented herein with additional explanations.

The theory of possibility of meteorite impact and cataclysmic effects has gained momentum owing to the fact that the major asteroid impact that occurred during geological history - 65 million years ago near Mexico that has contributed immensely towards elimination of nearly 30 % of marine and continental biota. From estimates of the terrestrial cratering rate, the frequency of similar event on Earth is of the order of one every 50-100 million years. Smaller, but still significant impact events, occur on shorter time scales and will affect the terrestrial climate and biosphere to varying degrees. Even the formation of impact craters as small as 20km could produce destructions similar to that of a nuclear winter. Such impacts occur on Earth with a frequency of two or three every million years. The most recent known structure in this size range is Zhamanshin in Kazaksthan, with a diameter of 15km. Impacts of this scale are not likely to have a serious effect upon the biosphere and cause mass extinctions. The most fragile component of the present environment however, is modern human civilization. Dependency of humans on an organized and technologically complex infrastructure for its survival makes the human community highly vulnerable to meteoritic impact related disasters as such events could seriously disrupt infrastructure facilities.

Impacts may also induce chemical changes in the atmosphere. These are related to the vaporization of the impacting body and a portion of the target. Considering only the contribution from the impacting body, recent calculations indicate that even relatively small impacting bodies, < 0.5 km in diameter that produce impact craters on the scale of 10 km in diameter, would inject 5 times more sulphur into the stratosphere than its present content. Larger impact events occurring on the time-scale of a million years will inject enough sulphur to produce a drop in temperature of several degrees and a major climatic shift. There are additional effects on atmospheric chemistry, including destruction of the ozone layer from shock heating atmospheric nitrogen and the injection of fluorides from the vaporized impacting body. The threshold for these effects is on a time scale of 100,000's of years.

Tsunamis would be generated if the meteorite impact occurs in oceans. In geological history also, tsunami events resulting from meteorite impact in marine regions are recorded. The 1960 tsunami, generated by a magnitude 8.6 Chilean earthquake, is thought to have been the largest this century. An impact-generated tsunami 10 times more powerful will occur with a typical recurrence time of a few 1000 years. The Tunguska event in 1908 was due to the atmospheric explosion of a relatively small, approximately few tens of meters body at an altitude of 10 km. The energy released has been estimated to be 10-100 megatons TNT equivalent. Although the air blast resulted in the devastation of 2000 sq. km of Siberian forest, there was no loss of human life due to the very sparse population. Events such as Tunguska occur on a time scale of 100's of years in which case, human race is at risk.

Quicksand

Quicksand is one of the potential dangers of natural process of geological phenomenon. Regions of quicksand develop in undisturbed marshy and swampy regions owing to many reasons including surface tension, bonding between grains due to the presence of organic matter, leafy materials that float over water over which sediments are deposited and left undisturbed. It was commonly believed that the behavior of quicksand was due solely to saturated or supersaturated suspensions of granules in water. Pressure from underground sources of water would separate and suspend the granular particles, reducing the friction between them. The stability of the colloidal quicksand is compromised by the presence of salt, increasing the likelihood of sand flocculation and the formation of the high viscosity regions of sediment responsible for quicksand's "trapping" power. Dry quicksand is loose sand whose bulk density is reduced by blowing air through it. It acts similar to regular quicksand, but it does not contain any water and does not operate on the same principle and yields quickly to weight or pressure. Until recently, the existence of dry quicksand was doubted. One region notorious for its quicksands is Morecambe Bay, England. As the bay is very broad and shallow, a person trapped by the quicksand would be exposed to the danger of the returning tide, which can come in rapidly. In India, many ephemeral rivers such as river Kaveri show the development of quicksand along the inner curves of meanders. It is also observed that owing to poor understanding, many deaths in such rivers due to quicksand are reported to be because of drowning under water.

Heat Wave

Heat wave is a geohazard that owes its allegiance largely to the burning of fossil fuels and rapid changes in atmospheric composition. Fatalities arising out of heat wave are linked to the result of anthropogenic intervention with natural climatological cycles through industrialization, landuse change, urbanization, deforestation etc. Heat wave is an extended summer resulting in human and other fatalities. Though a cyclic phenomenon, incidences of high fatalities by heat wave conditions in urban regions are linked to many causes such as population explosion, reduction in forest cover and unplanned growth of industrial and urban centers, etc. Implementing measures of reduction of contributing factors to heat wave and avoidance to exposure could potentially reduce the risks involved.

MITIGATION OF GEOLOGICAL HAZARDS

Mitigation embraces all measures taken to reduce both the effect of the hazard and the vulnerable conditions to it in order to reduce the scale of a future disaster. Therefore mitigation activities are not only focused on the hazard itself but also the elements exposed to the threat. Mitigation measures should be hazard specific and should focus on modifying the occurrence of the hazard. In addition, mitigation is also aimed at reducing the physical, economic and social vulnerability to threats and the underlying causes of vulnerability.

Although it is unrealistic to think that losses from geological hazards can be eliminated, there is no doubt that measures exist, and have been used, to reduce their impacts. The primary mitigation means are to avoid hazardous regions wherever possible and to build structures to resist their effects. Such actions require that we anticipate where the hazards are most likely to occur and to respond appropriately. In addition, under certain circumstances, it is possible to detect when a geological hazard is beginning to occur, which may make it possible to evacuate people from the threatened areas within days or hours of the event's onset and take steps to reduce destruction (Hamilton, 1997). Although the processes associated with geological hazards are distinct in many respects, there are many similarities as well, and their occurrences can, on occasion, be closely linked. Earthquakes and volcanic eruptions often trigger landslides and sometimes ice or water flows. Similarly, a cyclonic storm would render the coastal regions reel under flooding. Such generalizations could help plan mitigation efforts and most importantly, categorize regions according to potential vulnerability to each geohazard, namely "hazard zonation". Once different parts of country/state/district/ taluk/ block/panchayat/village are categorized to different levels of vulnerability to various geohazards, remedial and reclamation measures could be implemented and relief and rescue operations could be planned in advance.

HAZARD ZONATION

First step to be taken towards containment of impact of geohazard is hazard zonation at various scales. *Priori* knowledge about impending danger and its severity with reference to a particular region is important which can negate the surprise element. Hazard preparedness among common public and administrative personnel would largely reduce the potential impact and thus, hazard zonation is first exercise that is to be conducted. Geological hazard maps prepared by experts are used to provide information about the occurrence of past geological hazards and future potential occurrences.

Planners and policy makers in establishing national and regional development plans in order to minimize potential loss of human life and economic values due to geohazard events use them.

Owing to their unpredictability, many of the geohazards have caused damage especially in developing and underdeveloped countries. In view of uncontrolled and uneven socio-economic milieu, countries with weak and vulnerable social conditions experience aggravated damage by geohazards. One important aspect of mitigating damage is to make information about natural disasters readily available; not only to policy makers, planners, and professionals but to the general public as well. By doing so, communities will be better informed about devising measures to minimize the loss of lives and the economic impact of natural disasters. Natural hazard mapping and development of relevant databases are indispensable in such undertakings. Distribution of paper and digital versions of such maps is very important for people to gain such information as soon as possible (Kato et al. 2003). Constant monitoring and updating of hazard zonation maps is as essential as preparation of the map itself due to rapid population growth, urbanization and industrialization.

RELIEF AND RESCUE OPERATIONS

Since type or time or intensity or extent of geohazards that may occur in a specific place could not be predicted, overall preparedness to geohazards is essential, for which hazard zonation maps and creating awareness about geohazards could be helpful. Second most important step to be taken for mitigating geohazards and minimizing potential impacts are keeping ready a seamless relief and rescue operation system. For example, India has been traditionally considered as immune to volcanic and tsunami disasters, but recent occurrences of tsunami, eruption of mud volcanoes in Andaman region and increased incidences of mild tremors in India and adjoining coastal regions have raised the vulnerability of Indian sub-continent to geohazards hitherto unknown to Indian public. Although geohazards could neither be predicted nor prevented, a proper understanding of geohazards, assigning vulnerability indexes to Indian Territory or any other part of the World towards various geohazards and keeping the public prepared for potential geohazards could minimize life and property loss. Hazard mapping and risk assessment are two essential tasks that are yet to be undertaken in much of the Indian subcontinent. There is an urgent need for comprehensive vulnerability analysis for disaster-prone areas with additional data inputs of past disaster events, the socio-economic conditions of the population and inventories of major structures of public service and utility. Risk assessment and hazard mapping would delineate areas vulnerable to geology-related natural hazards and the frequency, intensity, impact, etc. of each hazard.

A major exercise of establishing integrated GIS based warning, relay and impact assessment system has to be undertaken immediately. Establishing an automated system for geohazard warning through a sophisticated network of satellite based data and GPS, forecasting infrastructure and an automated online GIS based tracking and broadcasting protocol to alert those concerned and evacuate public besides saving properties should follow it. This system has three components (*after* Ramkumar and Neelakantan, 2007).

First, entire continental land and adjoining coastal regions have to be mapped thematically through remote sensing supplemented with field data towards vulnerability for different geohazards based on various parameters such as (but not limited to) lithology, lineaments, structure, geomorphology, landuse/landcover pattern, coastal dynamics, slope, rainfall, drainage network, mean annual groundwater level, rate of groundwater recharge, sediment and soil types, forest cover, fracture density, thickness of weathered zone, depth to bedrock, erosion resistance of rock types, history of seismicity, flooding, cyclones, tsunamis, volcanic activity, coastal and river erosion, siltation of water bodies, infrastructure facilities and other socio-economic data, population etc. These parameters have to be assigned different weightage towards different geohazards that might be collated by GIS giving geohazard zonation output categorizing the study area in terms of low (20% vulnerability), moderate, high, very high and severe vulnerability (100% vulnerability). The output map has to be put in the Internet and made accessible to administrators, planners and all other stakeholders including general public. The output map has to be continuously updated with the help of remotely sensed data, GPS and GIS should there be any change in any of the parameters utilized, such as landuse pattern, geomorphology, coastal inundation, slope, population, etc.

Second, the output map should be linked with online GIS, which in turn should be connected with appropriate sensors planted in hazard prone areas. Once a major geohazard strikes in a particular area, the sensors would send signals through satellite to the ground control stations from where the critical information on potential zone of impact, probable lives and property under threat, types and means of relief and rescue operations to be undertaken would be relayed to the concerned officials. For example, had there been the system envisaged in this chapter in operation, the earthquake happened in Ausi Island of Indonesia and followed up initial coastal water withdrawal would have been registered in sensors and relayed to coastal dwellers and appropriate decision centers, through which the nature and potential impact of tsunami, that inundated Indian coasts could have been predicted within minutes with which precious lives could have been saved as there was a lime lag of few hours between earthquake occurrence at Ausi Island and arrival of tsunami waves to Indian coasts. Third, in the event of any geohazard related calamity, satellite data has to be obtained post-event and correlated with pre-existing output map for assessing magnitude of impact, type of rescue and relief operations to be carried out and also to decide on approach routes for transporting relief materials, etc. For example, in the event of a flooding, a casual glance on post-flood satellite imagery and its comparison with output map available online would give near-accurate estimate on croplands submerged, potential economic and life loss, etc, based on which resources and rescue personnel could be mobilized and sent to the affected area. Figure 1 shows how these components can be integrated and made operational. Establishing such an integrated system requires coordinated efforts of various academic and governmental agencies, besides creation of a wide network of appropriate sensors suitable for different geohazards, geohazard zonation of our country, etc.

This systems approach postulates that entire land and adjoining marine regions of a country be classified for geohazards vulnerability index based on all available data and made available in print and electronic format besides in a dedicated website. In the event of occurrence of a geohazards or any major natural calamity, the ground control sensors pickup the signals and automatically relay it to the research stations, satellite, alert the field inspection teams and also to appropriate decision centers. The research stations and field inspection teams, besides primary decision center personnel assess the impact, decide on relief and rescue systematics and start implementing it besides communicating the nature and extent of geohazards and its impact to appropriate higher levels of decision centers. Simple as it may sound to be, operationalization of this system warrants sophisticated automated systems, creation of synergy between command and control centers, human interface, convergence of data pickup and decryption, data analysis, communication media and feedback, etc.

SUGGESTIONS FOR MINIMIZING IMPACTS OF GEOHZARDS

Geohazards occur everywhere and no region of this earth is safe. With the opening of global economy and rapid strides in urban development and expansion of domestic, industrial, commercial, recreational and aesthetic land use in every available piece of land by human race, intervention into natural geological processes results, that lead to aggravation and finally, potential loss of life and property. While developmental activities could not be contained for the sake of non-intervention with geological processes, a well thought-out planning and developmental activities with minimal if not nil intervention with geological processes would thwart many a potential hazardous events. However, the very nature of geohazards being unpredictable it would be advisable to classify the land and water regions that are being utilized, according to vulnerability to various geohazards, and not to engage in potential catalytic activities. In addition, should there be any eventuality, being ready with contingent plans for relief and rescue operations would minimize loss.

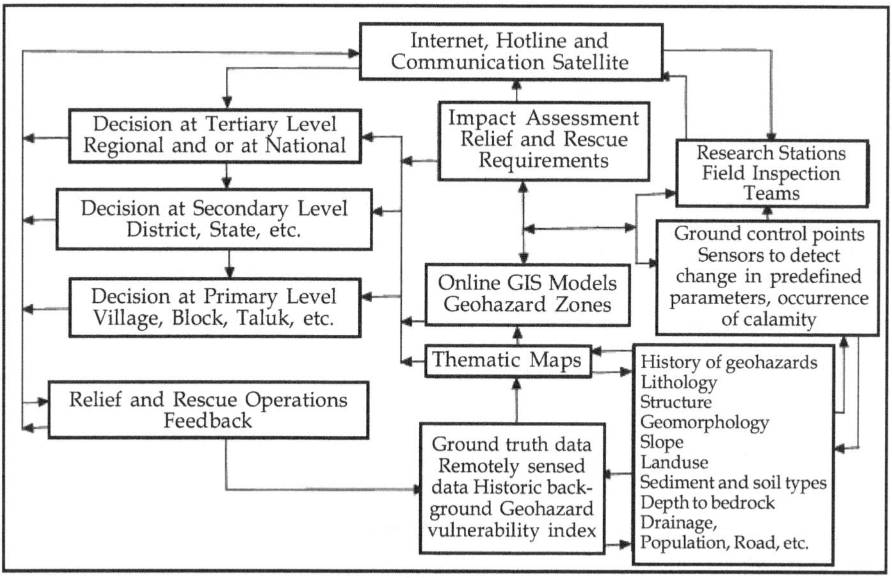

Figure 1. Conceptual model of geohazards mitigation, information dissemination and coordination system

Similar to the EIA reports (Environmental Impact Reports) being generated ahead of major infrastructural and other construction activities in India, potential geohazard reports have to be generated and the guidelines and suggestions made on the geohazard reports have to be implemented as in the case of EIA reports which have legal sanctity. In India, only strategy papers and policy guidelines are being generated and circulated currently, that too after the devastation caused by tsunami event of the year 2004. The geohazard report compiled after rigorous research and data collection for the proposed pipeline project in Azerbaijan (Geohazards report, 2002) is a classic study and similar studies have to be conducted. The suggestions made in such reports have to be legally binding.

Poor knowledge about geohazards is the most devastating hazard many countries including India face. For example, had it been known to the general public on the fateful day of 26th December'2004 that before advancing of huge tsunami waves, the sea used to retreat for about few meters to few kilometers, nobody would have ventured into beach to watch retreated sea to get perished under sudden deluge of tsunami waves. Similarly, earthquakes themselves normally won't kill lives, but the poor structures kill. When earthquake occurs, invariably aftershocks follow. Hence, strict implementation of regulations on construction of earthquake-resistant structures and avoidance of partially damaged structures is advisable. Better awareness on geohazards and measures for minimizing loss is essential.

Table 2. Nodal agencies and ministries responsible for disaster management

Disaster	Nodel Agency/Ministry
Natural disasters management (other than drought)	Ministry of Home affairs
Drought	Ministry of Agriculture
Air Accidents	Ministry of Civil aviation
Railway accidents	Ministry of Railways
Chemical disasters	Ministry of Environment and Forests
Biological disasters	Ministry of Health
Nuclear disasters	Department of Atomic energy

India has specific agency to deal with natural hazard and catastrophes. However, as shown in the Table 2, different catastrophes are dealt by different agencies/ministries. As geohazards bring in added complexity of various types of impacts, coordination among various agencies is essential. By this very nature, often relief and rescue efforts get delayed and or don't reach the needy at the hour of need, resulting in life and property loss. Over and above, technical Organizations, such as the India Meteorological Department (cyclone/earthquake), Central Water Commission (floods), Building and Material Promotion Council (construction laws), Bureau of Indian Standards (norms), Defence Research and Development Organization (nuclear/biological), Directorate of General Civil Defence provides specific technical support to coordination of disaster response and management functions. Such complexity in offering rescue and relief measures is often fraught with delays. Hence, it is suggested to establish a unified National Disaster Management Council, personnel drawn from all these agencies and assisted by appropriate machinery, infrastructure and personnel at appropriate levels - state, district, taluk, block, panchayat/village.

References

Arya, S., Karanth, A. and Agarwal, A., 2005 Hazards, disasters and your community: A primer for parliamentarians. Government of India. 62p.

Bandibas, J., Wakita, K. and Kato, H., 2003 Interactive Presentation of Geological Hazard Maps Using Geohazardview. Jour.Nat.Dis.Sci. v.25. pp.75-83.

Bhandari, R.K., 2006 The Indian landslide scenario, strategis, issues and action points. Keynote address delivered during India Disaster Management Congress, New Delhi, 29-30 November 2006. 19p.

ETC Pipeline ESIA., 2002 Geohazards report. 25p.

Government of India., 2005 Preventive/protection and mitigation from risk of tsunami. 34p.

Government of India., 2007 National disaster management guidelines.72p.

Government of India., 2004 Disaster management in India: A status report. 88p

Grieve, R.A.F., 1990, Impact cratering on the Earth, Sci.Amer. v.262. pp.66-73.

Hamilton, R., 1997 Report on early warning capabilities for geological hazards. IDNDR. 35p.

Integrated Global Observing Strategy., 2003 Geohazards theme report. 56p.

Jarraud, M., 2006 Drought monitoring and early warning: Concepts, progress and future challenges. Concept paper of World Meteorological Organization. 26p.

Kato, H., Wakita, K. and Bandibas, J.C., 2003 Eastern Asia geological hazards map: Paper and digital versions Jour.Nat.Dis.Sci. v.25. pp.65-74.

Ramkumar, M. and Neelakantan, R., 2007 GIS technology based geohazard zonation and advance warning system for geohazard mitigation and information dissemination towards relief and rescue operations. ICFAI Jour.Earth.Sci. v.1. pp.65-70.

www.wikipedia.com

❑❑❑

2

An Introduction to Tsunami and Characterization of Tsunamigenic Sediments with the Help of Microfossils

S.M.HUSSAIN
Department of Geology, University of Madras, Chennai - 600 025.
E-Mail: <smhussain7@hotmail.com>

This chapter reviews the published information on tsunami and attempts to characterize tsunamigenic sediments in terms of their microfossil content. Tsunamis are long-period waves generated by vertical displacements of marine watercolumn due to earthquakes, volcanic explosion and landslides. These waves can exceed heights of 25 meters (in shallow water) and lengths upto 100km with a velocity of more than 1000km/hour. Even at this magnitude, they are undetectable in deeper waters owing to very low wave height which in turn reduces significantly the response time, causing sudden havoc along coasts. They can cause flooding of coastal areas and affect areas thousands of kilometers away from their point of origin. These characteristics of tsunami imprint ensuing sediments with unique assemblage of microfossils.

INTRODUCTION

Shallow water progressive waves with very large wavelength and long duration caused by the rapid displacement of ocean water are called "tsunami". Tsunami is a sudden non-meteorologically induced impulse in water regardless of size. Tsunamis are formed by submarine tectonic activity more often due to earthquakes and less frequently due to volcanic eruption, landslides near the coast, meteor impact and nuclear explosions.

The velocity of the tsunami varies and depends on many parameters such as bathymetry, height of the originating wave, initial pressure changes in the sea, etc. and can travel at a speed of more than 1000 km/hr. At the place of origin (directly above the epicentre of submarine earthquake or any other causative factor), the wavelength may be upto few tens to several hundred kilometres but the wave height is often less than one metre only. Since the velocity is proportional to the depth of the water column, it gets reduced to few tens of kilometres as the tsunami approaches the coast. However, the wave height increases significantly reaching 100 m in extreme cases. As a result, the tsunami strikes the coast with massive force. The effect is colossal in islands with low relief. The tsunami waves appear as flooding with water rushing inside the land more violently and with great force towards a river course. In several cases, when the tsunami arrives, the water level in the sea may drop significantly and the waters receding several hundred metres come back with great speed drubbing the coastal lands doubly from both the directions. Though an event of short duration in terms of geological history, tsunami deposits, owing to their higher rate of deposition and other salient characteristics, find place in stratigraphic records. This chapter is an attempt to review the characteristics of tsunami and tsunamigenic sediments, to find out the ability of microfossil content of such sediments for characterization and to identify tsunami deposits in stratigraphic records.

HISTORICAL MAJOR TSUNAMI EVENTS

Tsunamis are among the most terrifying natural hazards known to man and were responsible for tremendous loss of life and property throughout history. Because of their destructiveness, tsunamis have notable impact on the human lives and social and economic security. Japan, which has one of the most populated coastal regions in the World and a long history of seismicity is also regularly affected by tsunami. There is also a history of tsunami destruction in Alaska, in the Hawaiian Islands in South America and elsewhere in the Pacific. The violent eruption and explosion of the volcano of Santorin, in the fifteenth Century B.C. generated a giant tsunami, which had destroyed most of the coastal Minoan settlements on the Aegean Sea islands. This event alone had acted as the catalyst for the decline of the advanced Minoan civilization. On 24th April 1771, 85m high tsunami waves struck the Japanese island of Ishigaki. They have hurled a 750ton piece of coral to a distance of 2.5 kilometers inland. Again in Japan, 27,000 people drowned in 1896, in a giant tsunami. A tsunami generated by a landslide struck at the island of Lanai in Hawaii about 105,000 years ago. The waves during this event had a hight of about 375m! An underwater volcanic eruption in island of Krakatau on August 26th - 27th 1883 triggered a wave 35 meters high that swept across the East Indies killing in excess of 36,000 people. The 1960 earthquake in Chile created tsunami waves that traveled more than 10,000 kilometers to Hilo, Hawaii. The 12 meters high water wall killed 61 people and destroyed many buildings.

Although not as frequent as in the Pacific Ocean, tsunamis generated in the Indian Ocean pose a great threat to all the countries of the region. The most vulnerable are: Indonesia, Thailand, India, Sri Lanka, Pakistan, Iran, Malaysia, Myanmar, Maldives, Somalia, Bangladesh, Kenya, Madagascar, Mauritius, Oman, Reunion Island (France), Seychelles, South Africa and Australia. During 1703 at Awa, Japan, more than 100,000 people lost lives due to a tsunami event. Tsunamis occur seldom in the Indian Ocean region, and in the last 300 years, this region had recorded only 13 tsunamis and 3 of them occurred in Andaman and Nicobar region for which the details of location of epicenter, death/damage caused etc. are not known. However, data on run-up heights indicate to the extent of 4 m in Port Blair with Nicobar recording very low (0.76 m) run-up height. Run up level is defined as the maximum elevation in land upto which it is inundated by seawater during tsunamis (Fig.1). Among these, the 1945 tsunami had a maximum run up of 13 m in Pakistan and resulted in death of 4000 people following an earthquake of magnitude 8.2 Ms in the Arabian Sea. Overall, the run-up levels varied from 1 to 13 m. In 1977, one of the strongest earthquakes of magnitude of Ms 8.1 struck west of Sumda Island in Indonesia, but there were no reports of casualties in India due to this tsunami. Apart from those listed, there are many events of tsunamis in the Indian Ocean that have not been properly documented. For example, villagers of Simeulue Island, off the coast of Sumatra, speak of a destructive tsunami in 1907 that had killed thousands of people.

On 26th December 2004, tsunami waves were triggered by a massive earthquake of magnitude 9.1 on Richter scale occurred at Sumatra due to subduction of Indian Plate beneath Burmese microplate below the Indian Ocean. The waves have traveled towards east and west with large wave length of several hundred kilometers and at a speed of about 700-800 km/hr, hit eastern India, Sumatra, Andaman and Nicobar islands and Thialand coasts, Sri Lanka and as far as Somalia coast in Africa in the west, inundated a vast stretch of coastal plains to the breadth of 1-3 km, incursed several kilometers along the rivers, streams and creeks, became the cause of death of over two lakh people and devastated the property worth of hundreds of millions of rupees. The eastern coastline of India bore the severe brunt of this event due to its short continental shelf region (an average of 32–45 km) than its western counterpart (about 175 km). Had the tsunami waves travelled directly onto the west coast of India (Arabian Sea), the damage would have been more, may be due to the amplitude or maximum height of a tidal wave (about 15 metres in the Gulf of Kutch and 13 metres in the Gulf of Cambay).

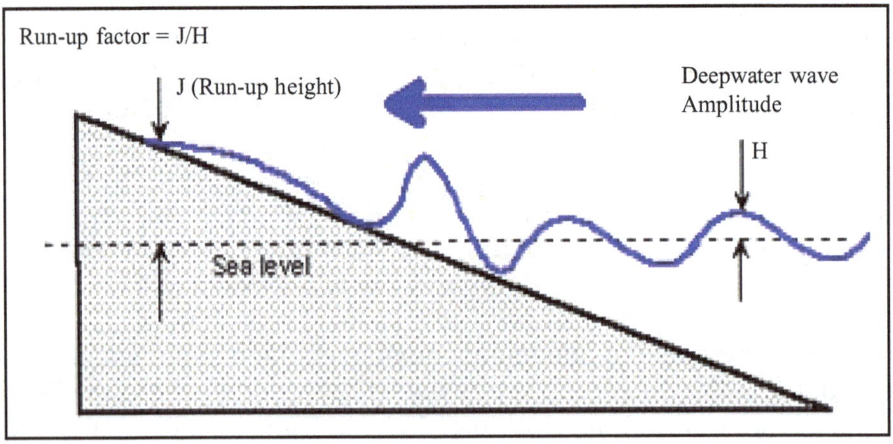

Figure 1. Inundation by tsunami waves and measurement of runup height

The Indian Ocean was considered relatively free of this catastrophic hazard, although, tsunami is possible on a lesser scale in the Bay of Bengal if massive earthquake occurred in the Andaman trench. Till this disaster struck, there was no strategy to combat the hazard in the South Asian countries. The need for early warning system in the north Indian Ocean was not taken seriously even though Geosciences Australia stressed the need for the establishment of an Indian Ocean warning system in the AUSGEO News 75 September 2004 edition. In a NASA and ISRO Conference on disasters held in June 2004 in Bangalore Roger Bilham from University of Colorado in his presentation *"Quantifying Tsunami Risk"* warned the possibility of a tsunami originating in the Indian Ocean. None expected that these warnings to become true so soon, that too with such a catastrophic level of intensity.

DETECTION OF TSUNAMI WAVES

Detecting tsunami is a very difficult task to do. At the point of origin, tsunami waves begin with a height of about 30-60 cm and look like nothing more than a gentle rise and fall of the sea surface. Tsunamis in deep water can have a wavelength greater than 500 kilometers and a period of about an hour. As they approach the shore, the wave height increases due to decrease in water depth and wavelength. In deep water, tsunamis travel at high speed. Destruction of roads, buildings and other man made structures is used as evidence for the erosional impact of tsunami. From past experience and concerted efforts, pressure sensors, GPS and other sophisticated satellite networks, significant improvements have been made towards detection of tsunami. Furthermore, communication and coordination systems are established between nation states that help forewarn impending tsunami dangers.

ORIGIN OF THE ASIAN TSUNAMI

The outer rigid layer of the earth is divided into 12 major lithospheric/tectonic plates. These plates are bounded by mid-oceanic ridges, subduction zones and transform faults. The boundaries are narrow deformation zones and accompanied by earthquake activity. The Indian plate is bounded by Sunda trench in the eastern side and Carlsberg ridge on western side. The earthquake of 26th December 2004 that occurred off the west coast of northern Sumatra took place at the interface between the Indian and Burma plates, where Burma plate has been referred by Andaman/Nicobar ridge that acts as a small tectonic plate. In this region, the Burma plate is characterized by significant strain partitioning due to oblique convergence of the India and Australia plates to the west and the Sunda and Eurasian plates to the east. It is a typical oceanic-oceanic convergent plate boundary where the Indian plate moving at a rate of 5.5 cm a year relative to the Burma plate colliding and getting subducted. Volcanism is common at such convergent boundaries. Such combination of two hazardous processes in the same region itself a dangerous condition to adjoining areas. Added to this, both of these processes could generate secondary hazards such as tsunami, landslide etc., raising the potential vulnerability still further. Furthermore, all these could wreck havoc in lands located thousands of kilometers away as demonstrated by tsunami of 2004, making the region highly vulnerable.

The earthquake of 26th December 2004 occurred off northwest of Sumatra in the vicinity of seismically active zone, close to Sunda Trench in the water depths of about 1300 m. Hypocenter of this earthquake is located relatively at shallow depth, about 30 km below the ocean floor (Yamanaka, 2004). It is the world's largest earthquake in the last 40 years since the 1964 Prince William Sound Earthquake in Alaska and is the largest in the history of Indian Ocean. More than one hundred aftershocks rocked the region. This extraordinary seismic event is consequent to the catastrophic release of strain, which has been accumulating for centuries as the oceanic lithosphere of the Indian plate is subducted beneath the Burma microplate at a rate of 6mm/year in a major undersea thrust. However, it took only 2 hours to realize that the seismic event is not an ordinary one. This high magnitude earthquake and its shallow epicenter have triggered tsunami in the northeast Indian Ocean. The tsunami moved with a speed of 800 km/hour with a curved path and travelled to about 4500 km within 7 hours inflicting damage to several countries of which, Indonesia, Sri Lanka and India are worst affected. The tsunami not only affected the South Asian countries but also Indian ocean island countries like Maldives and Seychelles and Somalia, located in the east coast of Africa causing lesser damage. It travelled further and went around the globe. 24 hours later, it reached southeastern coast of Australia with an amplitude of 40 cm.

The Disaster Control Research Centre (DCRC) of Tohoku University, Japan, based on numerical modelling has estimated the initial expected height of the tsunami as 2 to 3 m trough, then rising to a maximum height of 4 m. The shape of the wave front was crescent shaped almost parallel to the Sunda trench structure. The eastern front of the tsunami struck the Indonesian coast and Burmese coast within few minutes and in half hour reached the Nicobar coast followed by Andaman Islands and Sri Lanka. It reached the Indian Main land within 2 hrs inflicting colossal damage from Nagapattinam in the south to Chennai in the north (Photograph 1). The coast adjoining the Palk Strait and Gulf of Mannar were however protected by the Sri Lanka but the waves diffracted by it reached the Kanyakumari and South Kerala Coasts. The impact of the tsunami reduced further north of

Tamil Nadu and states like Andhra Pradesh and Orissa suffered lesser damage.

Photograph 1. Showing inundation of coastal land opposite Madras University in Chennai on 26th December 2004.

As many as 2,28,000 human lives were lost and several thousands still reported missing according to a statistics compiled after one month of this event. The Indonesia, due to proximity to the epicenter followed by Sri Lanka has experienced maximum loss of life and property. Despite being located close to the epicenter, the Andaman and Nicobar Islands experienced less number of life loss primarily due to very low population density. In the Indian mainland, the loss of life in Tamilnadu and Pondicherry is estimated as 8,264 with 426 villages affected. The data also show that the damage is colossal in Nagapattinam district followed by Kanyakumari and Cuddalore districts of Tamilnadu.

Nagapattinam, a coastal town, located about 400 km south of Chennai is the worst affected place in India due to this tsunami claiming more than 6000 death toll and extensive damage to the public and private property. It is believed that the dual wave effect (straight waves plus diffracted waves from Sri Lankan coast), gentle slope of continental shelf and gentle elevation of hinterland coupled with the presence of Uppanar river and Vedaraniyam canal in the southern side caused the deadliest impact in this district. Observations revealed that the run up level in the northern part of Nagapattinam near Light House is close to 4 m with a maximum inundation up to 1.1 km from the coast (Ram Mohan, 2006). Inundation of inland far from coastline was suggested to be because of the reason that the coastal lands were of gentle topography.

The post tsunami survey made by Marine Biological research team of Annamalai University, between Chennai and Nagapattinam had brought out startling facts. According to the results of this survey, the impacts were different in these two regions in terms of water quality, microbiology, plankton and benthos. In general, the tsunami waves altered the morphology of estuaries and backwaters. There were variations of impacts between water quality parameters of water samples collected at 5 and 10 km of the offshore waters, signifying bathymetry controlled variation of impact of tsunami waves. The Cooum river water and bottom sediment collected prior to and post tsunami show abnormal high concentrations of nitrogen, may be because of contribution from tsunami waves as well as resuspension of bottom sediments. Similarly, the Nagapattinam, Cuddalore and Parangipettai waters were found to record higher levels of nutrients. In general, coastal water off Karaikal was found to record fairly lower level of nutrients compared to other areas. The microbial population was found to have only marginal differences between sediment and water column after tsunami. It is also brought out from this study that the bottom sediments recorded higher levels of alteration than water samples, may be because of rapid neutralization of aqueous media than solids (sediment). Entry of exotic organisms like *Salmonella* in the study area, which were hitherto unknown, are also recorded in post tsunami conditions. In general, it was found that the organisms previously inhabitants of 5-10 km offshore were recorded in regions between 0.5-1km offshore, indicating landward shifting of ecological niches by tsunami.

The net impact of tsunami is that significant amount of nutrients were added to the coastal waters from terrestrial sources, as well as from resuspended sediments which seem to have triggered the biological production and led to the promotion of *chlorophyll* levels. Added to this, the sewage mixed in coastal waters extended to distances even up to 10 km from shore which has promoted sewage feeding microbial organisms like E. *coli,* Faecal coliforms, *Salmonella,* etc. to these far off places which had restricted distribution only up to 3km offshore before tsunami. All these observations indicate that not only the coastal morphology and sediment characteristics, but also the water quality and ecological niches were affected by tsunami waves.

TSUNAMIGENIC SEDIMENTS AND MICROFOSSILS

It is only after 26th December 2004, the scientific community along with common public and policy makers in India realized the need for assessing the vulnerability of Indian coast for tsunamigenic damage. Department of Science and Technology, New Delhi has initiated an integrated study of the tsunami involving several scientific organisations with specific mandates to map the areas vulnerable to inundation, geomorphic changes, sediment characteristics of the tsunami deposits and changes in groundwater quality and micropaleontological investigations (Hussain *et al.* 2006a, b).

Foraminifers are predominantly marine organisms living on the sea floor, or floating in the water column. Foraminifers have been, for the past several centuries, interesting objects to study geologically. These protozoans have adapted to various niches on the seafloor, in estuaries, salt marshes and in other marine marginal water bodies. Their smaller size and abundance in sediments of the present day and geologic past has resulted in foraminifera becoming the most used fossil group for determining the age and depositional environment.

Benthic foraminifera are useful in environmental studies because they are easily acquired, live primarily in the uppermost centimeters of sediments and are very abundant in marine and estuarine habitats. Recently benthic foraminifera have been used as powerful tools for tracking transgression and regression events in the lagoon and open sea. Like storm events, tsunamis can also be recognized in the stratigraphical record from the identification of foraminifera. However, on the basis of foraminifera alone, it is difficult to distinguish between storm deposits and tsunami deposits. Other criteria are needed to be considered, including stratigraphical context, spatial distribution of coarse-grained sediments and sedimentological characteristics. Clague *et al.* (1999) analyzed foraminifera in a core collected from Catala Lake on the west coast of Canada. They identified a gravel layer in the core, which they suggested was deposited in 1700 AD by a large tsunami caused by a historically documented earthquake. The tsunami layer itself is barren of foraminifera. Below the layer, the fauna consist of an almost monospecific assemblages of *Miliammina fusca*. Much more diverse marsh assemblages appear in the lake after deposition of the tsunami layer, including *Miliammina fusca, Trochomina rotaliformis, Trochomina inflata and Trochomina ochracea.* Based on this observation, the authors suggest that this assemblage reflect higher salinity conditions following deepening of the connection with the open sea by the tsunami and, possibly, by co-seismic subsidence. The tsunami sandy layer from Meares Island, Canada contains foraminifera transported from deeper water. If the sand layer was deposited during a storm, it is difficult to explain the change in height of the former marsh surface. A seismic event is likely to have caused the sudden drop of the marsh surface. This makes it plausible that the sand layer is tsunamigenic in origin.

Nelson *et al.* (1996 a, b) summarized the stratigraphic, sedimentological and paleontological evidences for the occurrence of tsunami deposits synchronous with episodes of co-seismic submergence at the Cascadia subduction zone of western North America. They drew attention to the occurrence of thin (marine) sand layers overlain by intertidal muds and underlain by buried salt marsh surfaces. Nelson *et al.* (1996) argued that the lack of similar sand beds within beds of peat or mud at these protected sites indicated that sand deposition was restricted to times of rapid submergence and hence unrelated to storm waves.

Relatively few investigations have been undertaken on the microfossils associated with tsunami deposits. Hemphill – Haley (1995a, b, 1996) has described various diatom assemblages contained within tsunami deposits along the Pacific coastline of Oregon, Washington and British Columbia. In her papers, she has reported a variety of brackish-marine diatoms within tsunami deposits that could be used not only to identify tsunami deposits but also to estimate the greatest inland extent of tsunami inundation owing to the fact that certain assemblages occur beyond the landward limit of tsunami-deposited sands and silts. Furthermore, she noted that in certain coastal freshwater lakes located beyond the reach of storms, the presence of marine planktonic and neritic plankton can be used to identify former episodes of tsunami inundation.

Dawson *et al.* (1996), in a investigation of diatom assemblages contained within the Grand Banks tsunami deposit, showed that the major diatom species present was *Paralia sulcata* and that the majority of individuals were broken. The most detailed investigation of foraminifera contained within tsunami deposits has been undertaken by Dominey-Howes et al. (1999) in western Crete. They demonstrated that tsunami deposits dating to the first centuary AD contain a wide variety of species. In particular, the presence of certain species indicated transport from deep water offshore, and this observation was used to support the concept of a former tsunami having taken place. Study of foraminifera contained within tsunami deposits is in its infancy and its true potential remains to be discovered.

Ostracods are bivalved tiny crustaceans; inhabit almost all types of aquatic environments including intertidal to abyssal depths, hot water or sulphurous springs. They are also found in freshwater and habitats ranging from temporary ponds, rice fields to rivers and deep lakes. Brackish water environments such as coastal saline lakes, coastal lagoons, and open estuaries also support a diverse range of these fauna.

Owing to these utilities, a specific study documenting presence of foraminifers and ostracodes of sediments of coastal marine regions along the stretch of Ennore to Nagaptnam in Tamil Nadu and in Andaman and Nicobar islands was conducted. The study had revealed that the faunal abundance and population is more in post tsunami sediments than the pre-tsunami samples in the offshore region of the sample collection (shallow depths). It is presumed that due to the tsunami activities, large amount of sediment might have been transported from the inner to outer shelf region and deposited near the shore regions. All the ostracodes recorded in these tsunami sediments are benthic forms and occur in neritic and upper bathyl zones. However, the samples were collected from beaches, creeks and river mouth upstreams. This observation supports the fact that the tsunami sediments have been brought from the inner to outer shelf region and deposited in these marginal marine water bodies.

CONCLUSIONS

The occurrence of the large and catastrophic events like tsunami cannot be predicted. However, certain issues related to these events may be addressed for the welfare of the mankind in future. Establishment of early tsunami warning system, creating better awareness on potential vulnerabilities and methods of relief and rescue operations among common public and official machinary may be helpful for the speedy evacuation of the people from the vulnerable areas. Afforestation with mangrove vegetation along the coastline at suitable sites may save the life and property of the coastal community. The strict implementation of coastal zone regulations could minimize the loss on the coastal life and infrastructure. Vulnerable locations have to be demarcated and various alternate arrangements have to be made for the coastal community. The research work incorporating all aspects of geology, environment, sociology and biology related to tsunami and various strategies prepared shall be utilized to mitigate the loss of life and property, which is the ultimate goal of any scientific study. Present study of documentation of microfossils from sediments of in pre and post tsunami event unequivocally had indicated the transport of deeper marine conditions as well as ecological niches to shallower and onnald regions due to tsunami waves and thus it is concluded that microfossil occurrence and abundance can be utilized to recognize tsunami events in stratigraphic record.

References

Clague, J.J., Hutchinson, I., Mathewes, R.W. and Patterson, R.T., 1999 Evidence for late Holocene tsunamis at Catala Lake, British Colombia. Jour.Coast.Res. v.15. pp.45-60.

Dawson, A.G., Shi, S., Dawson, S., Takahasmi, T. and Shuto, N., 1996a Coastal sedimentation associated with the June 2nd and 3rd 1994 Tsunami in Rajegwesi, Java. Quat.Sci.Rev. v.15. pp.901 - 912.

Dawson, A.G., Smith, D.E., Ruffman, A. and Shi, S., 1996b The diatom biostratigraphy of tsunami sediments: Examples from recent and middle Holocene events. Phys.Chem.Earth. v.21. pp.87 - 92.

Dominey-Howes, D.T.M., Dawson, A.G. and Smith, D.E., 1999 Late Holocene coastal tectonics at Falasarma, Western Crete: A sedimentary study. Geol.Soc.London.Spl.Publ. v.146. pp.343 - 352.

Hemphill-Hally, E., 1995a Diatom evidence for Earthquake induced subsidence and tsunami 300 years ago in southern coastal Washington. Geol.Soc.Amer.Bull. v.107. pp.367–378.

Hemphill-Hally, E., 1995b Intertidal Diatom from Willapa Bay, Washington: Application to studies of small-scale sea level changes. Northwest. Sci. v.69. pp.29–45.

Hemphill-Hally, E., 1996 Diatom as an aid in identifying late Holocene tsunami deposits. The Holocene. v.6. pp.439 - 448.

Hussain, S.M., Krishnamurthy, R., Suresh Gandhi, M., Ilayaraja, K., Ganesan, P. and Mohan, S.P., 2006a Micropaleontological investigations of tsunamigemnic sediments of Andaman Islands. Curr. Sci. v.91. pp.1655 - 1667.

Hussain, S.M., Suresh Gandhi, M., Krishnamurthy, R. and Nagendra, R., 2006b Micropaleontological investigations of tsunami sediments of Tamil Nadu and Andaman Islands. *In:* Rajamanickam, G.V., Subramaniyam, B.R., Baba, M., Ramesh, R., Elango, L. and Prithviraj, M. (Eds.). 26th December, 2004 Tsunami - Causes, effects, remedial measures, Pre and post tsunami disaster management: A geoscientific perspective. New Academic Publishers, New Delhi. pp.83-115.

Nagendra, R., Kamal Kanna, B.V., Sajith, C., Gargi Sen, Reddy, A.N. and Srinivasalu, S., 2005 A record of formainiferal assemblage in tsunami sediments along Nagapattinam coast, Tamil Nadu. Cur. Sci. v. 89. pp.1947-1952.

Nelson, A.R., Shennan, I., and Lono, A.J., 1996 Identifying coseismic subsidence in tidal wetland stratigraphic sequences at the Cascadia subduction zone of western North America. Jour.Geophys.Res. v.101. pp.6115 - 6135.

Ram Mohan, V., 2006 Inundation of seawater due to the Indian Ocean Tunami along Markkanam and Kovalam. In: Rajamanickam, G.V., Subramaniyam, B.R., Baba, M., Ramesh, R., Elango, L. and Prithviraj, M. (Eds.). 26th December, 2004 Tsunami - Causes, effects, remedial measures, Pre and post tsunami disaster management: A geoscientific perspective. New Academic Publishers, New Delhi, pp.12-36.

Yamanaka, Y. 2004 04/12/26 Off West Coast of Northern Sumatra Coast of Northern Sumatra, (http://www.eri.u.tokyo.ac.jp/sanchu/seis.note/2004/EICI61 e.html).

Yassini, I. and Jones, B. G., 1995 Recent Foraminifera and Ostracoda from Estuarine and Shelf Environments on the Southeastern Coast of Australia, University of Wollongong Press, Wollongong. 484p.

□□□

3

History and Geotectonics of Tsunami with Special Reference to the Indian Ocean

G.MANIMARAN
School of Tectonics, Department of Geology, V.O.C. College
Tuticorin - 628 008. E-mail: <acugemmani@yahoo.co.in>

A review on historic occurrences of tsunamis in the Indian Ocean and their associated characteristics followed by detailed characterization of genetic causes of recent tsunami are presented in this chapter. Although submarine earthquakes occur very frequently, resultant tsunami and its strength relied on many factors including transformation of energy into the ocean water, area of rupturing of ocean floor, duration of rupturing, accompanying submarine slides, kind of material at focus of earthquakes and type of faulting. The direction of tsunami wave propagation plays the decisive role in deciding sites of impacts. Perfect blend of all these parameters is required to make a submarine earthquake to generate devastating tsunami, which is relatively rare in geological history. This is the reason why although there were as many as 77 tsunamis that occurred in the Indian Ocean region during historic past, the recent one of 26th December 2004 is deadly and rare geological occurrence. However, aftermath effects in terms of volcanic eruptions, coastal high wave activities and change in climatic cycles are recorded since then and require constant scientific vigil.

INTRODUCTION

Tsunami is defined as a sea wave of local or distant origin that results from large-scale seafloor displacement associated with major earthquake or submarine slide or submarine volcanic explosion. Many a times it is incorrectly described as tidal wave or seismic sea wave, as both of these terms mislead in regard to its mechanism of generation. In Tamil Sangam literature, tsunami is cited as "Azhipperalai". In South America, it is termed as "Maremoto". Other names for tsunami include "Flutwellen", "Vloedgolven" "raz de mare" "Vagues Sismique" and Kai e'e. Prevalence of specific terms for tsunami in many such regions and languages itself stands testimony to the fact that tsunami is known to ancient human populations and that tsunami occurred frequently the historics past.

A tsunami is a system of waves formed in deep sea as result of sudden and large-scale disturbance of sealevel (Fig.1). Consequently, water being aqueous medium, attempts to return to normalcy at the earliest. Waves are generated in the process of restoration of sealevel to equilibrium through a series of oscillations, which in turn, are tsunami waves. Such a phenomenon can occur usually due to the onset of earthquakes under the sea. But in some cases, tsunami can also occur without earthquakes. For example, volcanic eruptions, displacement of marine sediments, landslides near the coast, nuclear explosions in the sea and and meteor impact from space can also generate destructive tsunami.

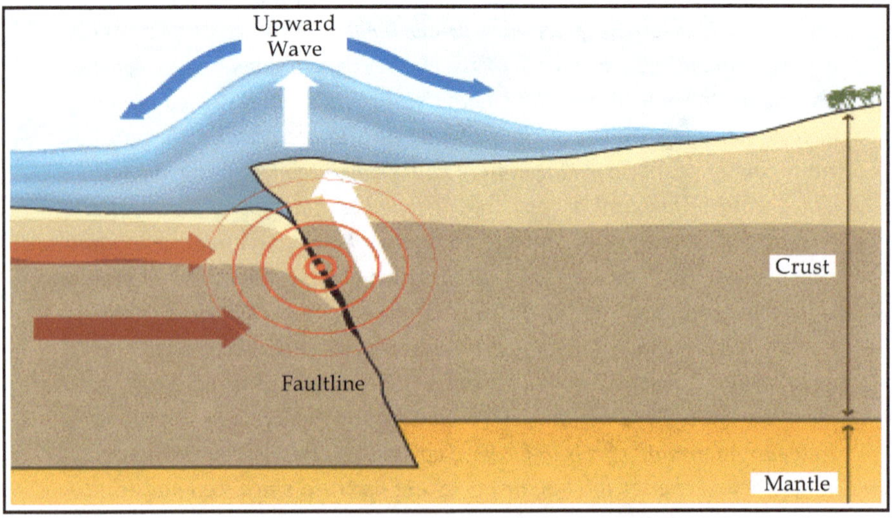

Figure 1. Diagram depicting faulting of seafloor and genesis of tsunami

HISTORY AND DISTRIBUTION OF TSUNAMI EVENTS

Out of 1244 tsunamis recorded worldover until the year 2007, nearly 80% originated from Pacific Ocean among which 82% were due to submarine earthquakes. Rest of 12% is from Atlantic Ocean and 6% from Indian Ocean (Table 1, modified after Ramasamy and Kumanan, 2005). This data suggests that the Indian Ocean is the least affected region when compared with other oceanic regions.

Table 1. Statistics on tsunami events of the past

Year	Magnitude	Pacific	Atlantic	Indian Ocean	Total
AD-0-BC	6-8	-	8	3	11
0-500 AD	5.5-8	-	11	2	13
501-1000 AD	6.4-8.6	13	8	1	22
1001-1500 AD	6.2-8.4	19	13	-	32
1500-2000 AD	3-9.5	971	103	66	1140
2001-2007 AD	6.4-9.3	20	1	5	26
Total	**1023**	**144**	**77**	**1244**	

Among the tsunamis generated in Pacific Ocean, 17% were reported in or near Japan; 15% from South America, 13% from New Guinea Solomon Islands, 11% from Indonesia, 10% from Kuril Islands and Kamchatka, 10% from Mexico and Central America, 9% from Philippines, 7% each from New Zealand and Tonga and Alaska, West Coasts of Canada and the United States and the balance of 3% from Hawaii. From these data, it could be inferred that even though the Pacific Ocean is considered to be tsunami prone, different countries experience varied impact of tsunamis. The country that is located at or near major tectonic and volcanic terrain undergoes heightened threats of tsunami whereas other countries experience least threat.

PHYSICS OF TSUNAMI

In terms of physics, transfer of large quantum of energy from earth's crust to ocean water within a short span of time and dissipation of the transferred energy by the water column constitutes a tsunami. A tsunami is a series of Ocean waves generated by large disturbances of sea water column (Large energy transfer from earth crust to Ocean water). Tsunami waves are characterized by long wave lengths (in excess of 100-500km), long wave periods (upto an hour) but very low height (few tens of centimeters) in deeper waters and very high (upto

375m on record) in shallow waters. Increase in wave height is proportional to decrease in bathymetry with concomitant decelaration of velocity. It is explained such that shallow-water waves move at a speed that is equal to the square root of the product of the acceleration of gravity (9.8m/s^2) and the water depth. For a 4km deep Ocean, the predicted wave speed is about 200m/s and a wave packet can cross a 5000 km Ocean in about 7 hours; i.e., the wave can travel at the speed of a jet plane crossing the entire Pacific Ocean from one side to other side in less than a day. Because the rate at which a wave loses it energy is inversely related to its wavelength, tsunamis not only propagate at high speeds, they can also travel great distances with limited energy loss. As tsunami waves are water waves in which the restoring force is gravity and the wavelength is greater than the Ocean depth, unlike shorter wavelength disturbances, the fluid motion extends throughout the water column.

Tsunami waves travel outward in all directions from the epicenter. Generally, the energy is propagated in a direction orthogonal to the strike direction of fracture. There are two interacting processes that allow these long gravity waves to propagate. The first is the slope of the sea surface, which creates a horizontal pressure force. The second is the piling up as water moves with different speeds in the direction that the waveform is moving. When these two processes have the right relationship in time, they create propagating waves. This may be the reason why not all submarine earthquakes or other causative natural processes become rarely tsunamigenic.

As the tsunami leaves the deepwater of the open ocean and travels into the shallower water near the coast, it slows down. The tsunami's energy flux, which is dependent on both its wave speed and wave height, remains nearly constant. The speed of tsunami diminishes as it travels into shallow water, its height grows, owing to the piling up of water. Because of this shoaling effect, a tsunami, imperceptible at sea, may grow to be several meters or more in height near the coast. When it finally reaches the coast, a tsunami may appear as a rapidly rising or falling tide as a series of breaking of waves. May be because of this characteristic of tsunami, it gets noticed only when it strikes and it catches its victims unawares.

As the tsunami crosses the deep ocean, its wavelength may be hundred kilometers or more and its wave height would be only a few tens of a centimeters. The waves cannot be felt aboard ships nor can they be seen from the air in the open ocean. The regional variation of ocean depth cause partial or even complete reflections and waves can diffract around Islands and coastlines. As tsunami attacks the shoreline, the wave energy is forced to be spent in a region of much shorter distance and shallower depth resulting in churning-up of the entire water column and life-threatening waves. Thus the tsunami waves have great erosional and impounding potential, capable of inundating or flooding, hundreds of meters inland, crushing of homes and coastal structures and also depositing huge, lengthy layer of tsunamite deposits on coastal plains.

Tsunami may reach a maximum vertical height onshore above sea level, often called a run up height of 10, 20 and even 30m and beyond. The first wave of tsunami is not likely to be the biggest. Actual tsunami wave heights can be measured from the amplitude of the wave signals seen on tide gauge instruments. Inundation distance is defined as the maximum horizontal distance inland that a tsunami penetrates. Inundation height is a measured height at the maximum inundation distance with reference to the normal sea level. The run-up height is not necessarily the same as the inundation height near the shore. Energy scales of the tsunami in terms of tsunami intensity (i) and tsunami magnitude (Mt) are proposed by various tsunami experts and are detailed herein.

1. Tsunami Intensity (i)

$i = \log^2 (\sqrt{2}h)$ *where* h = mean tsunami height. Mean tsunami height $= 1/\sqrt{2}$ times the maximum height of Tsunami, i = Maximum tsunami intensity near the epicenter.

2. Intensity of the Tsunami

$\text{Log}^2 (2^{1/2} \times h)$ *where* h = maximum run-up height in meters. The range is 0.5 to 10.

3. Tsunami Magnitude (Imamura - Lida Scale)

$m = \log^2 h$; *where* h= maximum run-up height in meters.

Tsunami magnitude (Mt) predictive equation was proposed by Abe (1981) for regional tsunami. $Mt = \log H + \log \Delta + 5.8$ *where* H = Maximum amplitude in tide gauge in meters, Δ is the actual distance in km.

For Trans - Pacific tsunami, $Mt = \log H + C + 9.1$; *where* H = Maximum amplitude in tide gauge in meters and C is the distance factor depending on the combinations of the source and the observation point.

$Mt = \log H^2 + \log X + 5.55$ *where* H^2: Maximum crest to trough amplitude on tide gauge record in meter. X: Distance from epicenter to station along the shortest oceanic path in kilometers (Abe, 1981).

HISTORY OF TSUNAMI IN THE INDIAN OCEAN

The phenomenon of tsunami is known from ancient times. Research on historical occurrences of tsunami may prove helpful in analyzing its frequency of occurrence. It is known from the manuscript written by Plato, wherein he has described a giant sea wave that destroyed the town of Amnisos on Crete in the year 1628 B.C. due to the explosive series of eruptions from the great volcano Santorini in the Aegean Sea. A reference on the engulfment of Dwarka, a city in Gujarat coast about 1500 - 1400 B.C. can be traced in Indian Epic Mahabharatha. According to Iyengar and Radhakrishna (2005), the drowning of the city Dwarka can be attributed to a great tsunami of that time.

Long before the Christian era, the ancient Tamil had nurtured the three Tamil Sangams and their rich language flourished in their academics. The Early Sangam had operated at Thenmadurai, South of Kanyakumari between 300 B.C. and 50 B.C., The Middle Sangam had operated at Kapaadapuram, a coastal town, possibly near NE of Kanyakumari between 50 B.C. and 100 A.D. Late Sangam had functioned between 100 A.D. and 300 A.D. at Madurai, a city far away from the coast. The Sangam literary works namely, Kalithogai, Kurunthogai and Cilappathikaram refer the lost coastal cities of Thenmadurai and Kapaadapuram due to past tsunami attacks possibly around at 50 B.C. and 100 A.D. respectively (In Tamil, Perooly means Tsunami event). Cilappathikaram, an epic more than 1700 years old (Appadurai, 1941; Karthigesu Sivathamby, 1986) states that *"Pahruli Atrudan Panmalai adukathu Kumari Kodum Kodunkadal Kolla"*. It means the ferocious sea has engulfed the river Pahruli along with mountainous terrain upto Kumari". Until recent history, the tourist spot located south of Cape Comerin namely Swamy Vivekananda Memorial Rock had had a continuous mountainous region from main land for about few kilometers long and wide, but remains as a separate island now and stands testimony to similar marine transgressions that took place during sangam periods.

The Poompuhar, located at north of Nagapattinam, a east coast city of Tamilnadu, also an ancient old harbor under the rule of Chola Kingdom was swallowed by a "Azhipperalai" that occurred around 1500 years B.P. (i.e., 500 A.D.). Ancient scriptures and literature in the library of the Thondaiman Kingdom in Pudukottai of Tamilnadu provide evidence for tsunami of Indian Ocean that struck Nagapattinam in 900 AD; destroyed a Buddha Monastery, washed several temples and killed hundreds of people during the reign of Raja Raja Chozhan. In Indian History, the records on earthquake-induced tsunamis are available from November 326 B.C., when a tsunami hit the Indus Delta/Kutch region. It destroyed the Macedonian fleet owned by Alexander the Great. During the year 1524, a tsunami struck the west coast of India due to earthquake occurred at Dabhol, a coastal town of Maharashtra. In 1618 A.D. a tsunami accompanied with hurricane occurred at near off the coast of Bombay (Mumbai).

Apart from these eight tsunami events described in preceding paragraph, National Geophysical Data Centre (NGDC) and Geological Survey of India have documented a total of 69 tsunami events, and are tabulated in chronological order (Table 2). Perusal of the ancient literature, epics and data from NGDC and GSI indicate all these tsunamis were generated from three regions around India namely, The Kutch of Gujarat - Mekran (Pakistan) seismic belt, Java - Sumatra - Malaysia - Andaman - Nicobar and Myanmar Seismic belt and Srilanka - Kanyakumari and adjoining Indian Ocean belt (Manimaran and Renuga, 2007a, 2007b).

As many of the geological processes are cyclic in nature and the durations of which vary from few hours (tidal cycles) to few tens of millions of years (first order sealevel oscillations), the occurrence of major catastrophic events such as tsunami might also be studied in the light of historic perspective, primarily deduced from ancient literatures (of course with due diligence) and exact positioning and timing of such events have to be verified with the help of field investigations. Hence, it is felt that it is high time the tsunami events of the past have to be studied from various literatures of Indian languages and be checked in the coastal field to bridge the gap in understanding historical Indian Ocean tsunamis.

Table 2. Tsunami Events in the Indian Ocean and adjoining regions

Year	Mo	Dt.	Country and location	Latitude	Longitude	Depth (km) of source	Earthquake Magnitude (Ms)	Intensity of tsunami	Death toll due to tsunami
1750			Myanmar - Burma Coast	18.500	93.400			4.00	0
1762	4	12	Indian Ocean Northern Bay of Bengal	21.000	89.000				1
1770				5.000	102.000		7	0.50	0
1797	2	10	Indonesia -SW Sumatra	1.000	99.000		8	3.00	300
1816	4	29	Malaysia - Penang Island, Malacca Peninsula	5.383	100.250				0
1818	3	18	Indonesia - Bengkulu, Sumatra	3.767	102.267		7	1.50	1
1819	6	16	India -Rann of Kutch				7.5		
1833	11	24	Indonesia - SW Sumatra	3.500	102.200		8.2	2.50	0
1837	9		Indonesia - Banda Atjeh, Indonesia	5.500	96.000		7.2	0.50	0
1842	11	11	Indian Ocean - Northern Bay of Bengal	21.000	89.000				1
1843	1	5	Indonesia - SW Sumatra	1.500	98.000		7.2	2.00	1
1845	6	19	India - Rann of Kutch	23.583	68.367				1
1847	10	31	India - Little Nicobar Island	7.333	93.667				0
1852	11	11	Indonesia - Sibolga, Sumatra	1.700	98.800		6.8		0
1861	2	16		1.000	97.500		8.5	3.00	905
1861	3	9		0.300	99.370		7	2.00	1700
1861	4	26	Indonesia - SW	1.000	97.500		7	1.50	0
1861	9	26	Sumatra	1.500	100.000		6.5	1.50	0
1863	3	16	Indonesia - Java	6.133	106.750				0
1868	8	19	India - Andaman Islands	11.670	92.730				0
1874	5		Bangladesh Bhola district	Sunderbans of Bangladesh were affected					1200000
1881	12	31	Indian Ocean - Bay of Bengal: W of Car Nicobar Islands	9.000	92.000	15	7.9		0

Contd...

Table 2. Contd...

1882	1		Sri Lanka	8.570	81.230				0
1883	8	26		6.102	105.423			1.00	0
1883	8	27	Indonesia - Krakatau	6.102	105.423			5.00	36500
1884	2			6.102	105.423				0
1884			India - Western Bay of Bengal	Port Blair, Dublet (kolkatta) were affected.					
1885	7	29	Indonesia - Ajerbangis, Sumatra	0.200	99.383		6.8		0
1885	12	14	Indonesia - Banda Atjeh	5.500	96.000				0
1886	1	31	India - Bay of Bengal						0
1889	8	16	Indonesia - Java					1.00	0
1896	10	10		3.500	102.500		6.8		0
1907	1	4		2.000	94.500		7.6	2.00	400
1908	2	6	Indonesia - SW Sumatra	5.000	100.000	130	7.5	1.00	0
1909	6	3		2.500	101.500	40	7.7	1.00	0
1914	6	25	Indonesia	4.500	102.500		7.5		0
1917	1	21	Indonesia - Bali Sea	8.000	115.400	33	6.5		0
1921	9	11	Indonesia- S. Java Sea	11.000	111.000		7.5	2.00	0
1922	7	8	Indonesia - Lhoknga, Aceh	5.467	95.233				0
1926	6	28	Indonesia - SW Sumatra	1.500	99.500		6.7		0
1928	3	26	Indonesia - Krakatau	6.102	105.423				0
1930	3	17		6.100	105.400				0
1930	6	19	Indonesia - Java	5.600	105.300	33	6		0
1930	7	19		9.300	114.300	33	6.5		0
1931	9	25	Indonesia - SW Sumatra	5.000	102.700	3	7.5		0
1935	12	28		0.001	98.250	60	7.9		0
1936	8	23	Malaysia - Malay peninsula	6.000	95.000	7.3			0
1941	6	26	India - Andaman Sea	12.500	92.500	8.1	7.6		5000
1945	11	27	Baluchistan - Mekran Coast	24.500	63.000		8		4000
1948	6	2		5.500	94.000		6.5		0
1949	5	9	Malaysia - Malay peninsula	5.000	95.000		6.7		0
1955	5	17		6.500	94.000		7.2		0
1957	9	26	Indonesia - S.Java Sea	8.200	107.300		5.5		0
1958	4	22	Indonesia - SW Sumatra	4.500	104.000	33	6.5		0
1963	12	16	Indonesia – Java	6.200	105.400	64	6.5		0
1964	4	2	Indonesia - Off Northwest Coast	5.900	95.700	133	7		0
1967	4	12	Malaysia - Malay peninsula	5.500	97.300	58	7.5	1.50	0
1981	12	31	Indian Ocean - Bay of Bengal						0
1982	2	24	Indonesia - Java Trench	4.370	97.700	52	5.4		0
1983	11	30	Carlsberg ridge, Indian Ocean	Small Tsunami Struck Diego Garcia.					
1985	4	13	Indonesia – Bali island	9.200	114.200	99	6.2		0
1994	2	15	Indonesia – Sumatra	5.000	104.300	23	7		7
1994	6	2	Indonesia – Java	10.500	112.800	18	7.2	2.50	238
2000	6	4		4.721	102.087	33	8.0		103
2000	6	18	India - South Indian Ocean	13.800	97.450	10	7.9		0
2002	9	13	India - Andaman Island	13.040	93.070	21	6.7		0
2004	12	26	Indonesia - Off West Coast of	3.316	95.855	10	9.3	9.00	230000
2005	3	28	Sumatra	2.074	97.013	30	8.7		2000
2006	7	17	Indonesia - South Java Sea	9.222	107.320	34	7.7		600

(*Source:* National Geophysical Data Centre-Paula-dunbar@noaa.gov; GSI, 2005, Manimaran and Roy Chacko, 2006, Manimaran 2007)

GEOTECTONICS OF INDIAN OCEAN TSUNAMI

Although earthquakes were purported to be the primary cause of tsunami generation, not all earthquakes generate tsunamis. To translate a major earthquake event into a diastrous tsunami, the following conditions have to be satisfied.

- It should be a submarine earthquake with magnitude greater than 7.
- Vertical displacement of sea bottom surface either upwards or downwards.
- Large area of rupturing (hundreds to thousands of square kilometers)
- Mostly from shallow focus submarine earthquakes (depth < 70 km) and
- The released acoustic energy waves/gravity waves should be transmitted to the seawater.
- Other conditions such as bottom topography, duration of rupturing, etc have to be conducive.

In the light of these conditions, the tectonic framework of the Indian Ocean is detailed herein. The Indian plate gets subducted under Burma plate. In the Andaman regions, the Indian plate moves toward the N 23° E at the rate of about 5.6 cm/year relative to overriding Burma plate. The Andaman and Nicobar Islands parallel the arcuate plate boundary, separating the Indian and Sunda plates, at east of 90° East Ridge (Rajendran and Gupta, 1989). In the rupture zone of 28th March earthquake, the Australian plate moved towards northeast at a rate of about 6cm/year relative to Sunda plate. This had resulted in oblique convergence at the Sunda trench. The oblique motion is partitioned into thrust faulting, which occurs on the interface between Australia and Sunda plates.

The Indo-Australian plate has been pushing northward into Eurasia since 50 million years ago, which forcefully raise the Tibetan plateau and the Himalayan Mountains. As a result of this collision between India and Asia and due to buckling and shortening of Oceanic lithosphere since 8 million years ago (Stein and Okal, 1978), the Indo-Australian plate broke apart just south of the equator beneath Indian Ocean into separate Indian and Australian plates. This crustal break is located 140 km northwest of the epicenter of 28th March 2005 earthquake (Fig.2). The counter-clockwise rotation of new Australian plate in relation to the Indian plate has led to compression in the eastern boundary near Indonesia plate (convergence) and tension in the western part of the new plate (divergence) boundary (Gorden et al., 1990). The series of earthquakes that took place were on the eastern boundary of the new Australian Plate.

Strong Tsunami of Indian Ocean

On 26 December 2004, Indian Ocean tsunamigenic earthquake (Mw 9.3) occurred at north of Simeulue Island, off the western coast of northern Sumatra, Indonesia. The hypocenter of the main earthquake was at 3° 72" N; 95° 00" E at about 160 km west of Sumatra, at a depth of 30 km within the Indian plate. The epicenter of the earthquake was at a quadruple plate junction of Burmese microplate, Sunda plate Australian plate and Indian plate (Fig. 2). It is the second largest earthquake ever recorded. The death toll from this earthquake was more than 3,00,000 people. This disastrous tsunami waved up to 10 m high and swept the coastal regions of Indonesia, Thailand, Sri Lanka, South India, Somalia and Port Elizabeth in South Africa. Its impact was felt as far as 8000 km away from epicenter. The ground shaking was also felt in Bangladesh, India, Malaysia, Myanmar, Thailand, Singapore and the Maldives. Seiches were noticed on ponds in Tirunelveli, Tuticorin, Thanjavur and Cuddalore Districts of Tamilnadu, India. The height of the tsunami waves triggered by the Sumatra earthquake was 10m at Banda Aceh, Sumatra, 6 to 7 m in Andaman - Nicobar Islands and 1.5 to 3 m in coastal regions of Southern India. Marine water inundated up to distances of 0.1 - 4 km, depending upon the bathymetry and coastal geomorphology. The run up elevation in various places (in meters) are as follows: Pulicat 3.2, Pattinapakkam 2.7, Kovalam 4.3, Periakalapet 3.9, Puttupattinam 2.6, Devanaanpattinam 2.5, Parangipattinam 2.8, Tarangambadi 4.4, Nagapattinam 5.2, and Vedaranniyam 3.6 of east coast of India and at Sumatra 10 metres (Yamanaka, 2004; Harinarayana and Naoshi Hirata, 2005; Abe, 2005; Ramanujam et al., 2005, Narayana et al. 2005, EERI Survey, 2005, Manimaran and Roy Chacko, 2006 and Manimaran 2007). On 26th December 2004, about 1300 km long land along subduction plate boundary in the Sumatra and Andaman-Nicobar region slipped up to 20m on a gently dipping plane towards NE. In addition, uplift up to 5m uplift at Northern Sentiel Island, 1 m subsidence in the Andaman-Nicobar area, about 1 - 4 m southwestward horizontal movement of Andaman and Nicobar island

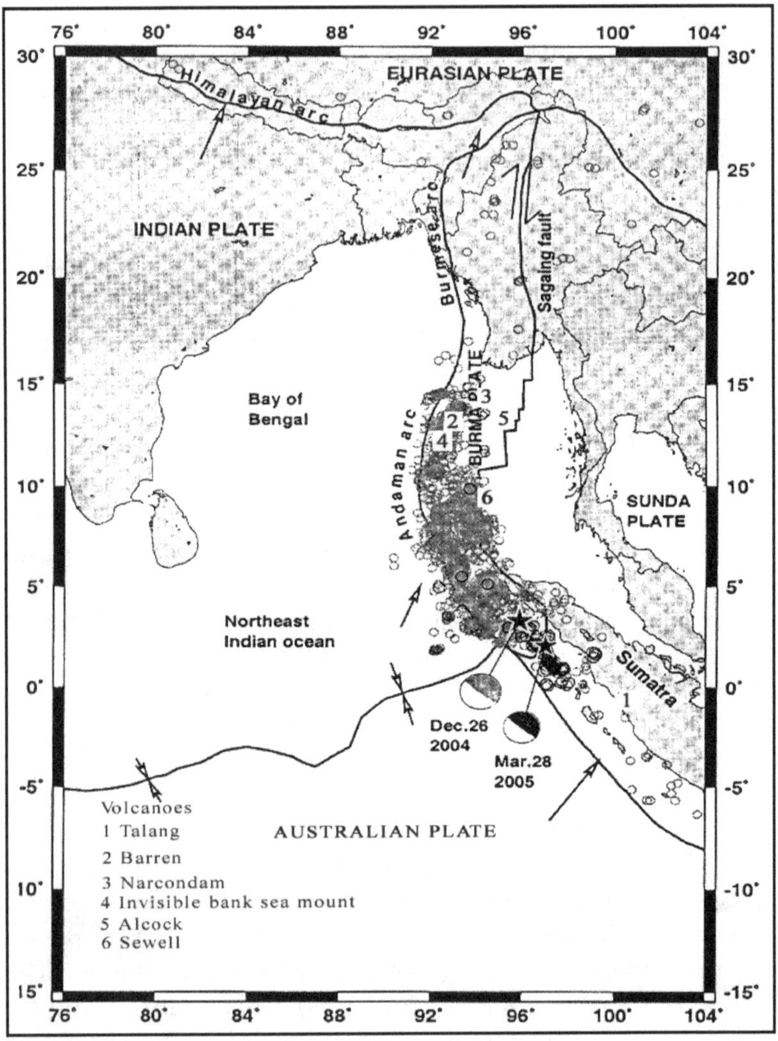

Figure 2. Location of 26th December, 2004 and 28th March 2005 Northern Sumatra earthquakes

These two earthquakes occurred near quadruple tectonic junction of India, Burma, Australia and Sunda Plates.

Location of the volcanoes

1. Talang
2. Barren
3. Norcondam
4. Invisible bank sea mount
5. Alcock and
6. Sewell are shown. (Modified after Rao and Hanumanthachary, 2005).

Eruptions in Talang, Barren and Norcondam Volcanoes were triggered by the aftershocks of 2004 and 2005 great earthquakes. (Manimaran and Ravindran, 2007).

regions and about 15 meters of horizontal displacement of Banda Aceh of Sumatra are reported. The peak vertical displacement of seafloor was about 5 meters (Bilham et al., 2005; Gahalaut and Chanda, 2005). Based on GPS measurements in September 2003 and February 2005, 6.49 ± 0.009 m horizontal slip with a 1.1m subsidence at Car Nicobar and 4.78 ± 0.008 m horizontal slip with uplift of 0.6 m at Diglipur, northern most Andaman were reported.

Weak Tsunami of Indian Ocean

Subsequently on 28th March 2005 a weak tsunamigenic but the second largest earthquake (M 8.7) in the world since 1964 occurred in Australian plate at Western shelf region of northern Sumatra (2.074° N and 97.013° E). Epicentre of this earthquake is located at about 160 km southeast of the main shock location of 26th December 2004 earthquake. It occurred along the interface between Australia and Sunda plates (Fig. 2). It resulted due to thrust faulting and caused extensive damage on Simeulue of northern Sumatra. Hundreds of buildings were leveled at Nias Island. Reported death exceeded 2000. It was felt in Indonesia, Malaysia, Singapore, Bangkok, Thailand and in south India. Seiches were observed on ponds of West Bengal, India. The peak vertical displacement of sea floor was 3 m. Although many governments considering the threat of a tsunami made evacuations in coastal regions of Thailand, Malaysia, Sri Lanka and India, no major tsunami was generated except a weak tsunami of 3 m high wave with maximum run up of 4.5 m that hit the Simeulue Island (Wikipedia, 2005).

Differentiation between Strong and Weak Tsunamis

The tsunami on 26th December 2004 was a direct consequence of the earthquake, which caused vertical movement (5 to 10m) of the sea floor over a huge area, like a giant wave machine. However, the 28th March 2005 earthquake (3m uplift of sea floor) did not generate destructive tsunami, although tide gauges in the Indian Ocean did show changes to the expected sea level (i.e. Wave heights of about (in meters) 0.50 at Panjang, Indonesia; 0.42 at Colombo, Sri Lanka; 0.23 at Cocos Island, Australia; 0.40 at Hanimadu; 0.22 at Male, 0.28 at Gan, Maldives; 0.35 at Port Louis, Mauritius, 0.35 at Ponte La Rue, Seyehelles and 0.30 at Tuticorin, India). Both earthquakes occurred under the seabed and are fairly comperable in magnitude. Also the fault mechanism is similar in nature. Failure of 28th March earthquake to generate a destructive tsunami might be due to one or more of the following reasons.

- The vertical displacement of the 28th March 2005 earthquake is comparatively small. Observations indicate about 1 meter of subsidence on the coast of Kepulan Banyak as well as 1 meter of uplift on the coast of Simeulue, unlike 26th December 2004 earthquake of 5 meters uplift of Burmese plate.

- The rupture pattern of fault zone and its speed and direction of propagation were different. The seismic data of indicate that there were two phases of rupturing of Sumatra region for about 1300 km long and 160 - 240 km wide area. The first phase involved the formation of rupture about 400 km long and 100 km wide and at 30 km beneath the seabed. The rupture proceeded at a speed of about 2 km/s, beginning off the coast of Banda Aceh and proceeding northward upto a distance of 400 km within a timespan of about 200 seconds. The second rupture generated ultra long period seismic waves (less energy in frequencies >5 Hz) due to slow and large dislocation for about 900 km in NNW direction up to northern Andaman Island. The maximum slip on the rupture was about 20 m. The total source duration was about 13 minutes (Abe, 2004; Stein and Okal, 2005). This lengthy period of rupturing might have resulted in tsunamigenic event. During 28th March 2005 earthquake, the rupture nucleated at 160 km southeast of epicenter of 2004 Sumatra earthquake and propagated 300 km towards southeast at a speed of 1.6 km/s with total rupturing duration of 3 minutes only. The rupture propagated mainly along Sunda trench axis. The maximum dislocation is about 3 meters (Yagi, 2005; Purnachandra Rao and Hanumanthachary, 2005).

- These rupturing patterns 2004 and 2005 earthquakes occurred independently within the respective Indian and Australian plates and suggest that there is a crustal break between them. The northerly procession of aftershocks of December 2004 Indian plate earthquake and southeasternly procession of aftershocks of March 2005 Australian plate earthquake were mostly confined within their plate boundaries (Fig.3). Near east of the 2005 Sumatra earthquake site, the Indo-Australian Crustal break trends NE to SW and along west, it trends ENE-WSW to E-W and joins the Indian Mid-Oceanic Ridge. The energy released during the March 2005 earthquake might have dissipated along Indo-Australian crustal break or might have been utilized for pushing the

Australian plate anticlockwise. Hence, the expected tsunami was not stronger enough to cause havoc.

- The strike-slip component to the Sunda Trench was more prominent than the vertical slip component of Sunda - Australian plate interface during March 2005 earthquake.
- The focus of the March 2005 Sumatra earthquake was located vertically below the chain of Islands and rupturing passed through both islands and sea bottom beds. This factor might have made non-availability of water column to be displaced and transformed into oscillatory tsunami waves. In addition, the fact that the December 2004 earthquake occurred in deep marine water (2 to 4 km) whereas the March 2005 earthquake occurred in a shelf region at water depths of less than 1 km, is suggestive of a situation that even if the March'2005 seismicity displaced watercolumn, owing to comparatively less bathymetry, the resultant tsunami might have been weak only.
- There was a significant difference between the two events in the primary direction of tsunami wave "focusing", which affected the distant or "far-field" impact of the tsunami. During December, tsunami energy was focused to the west, towards Sri Lanka and India and to the east, towards Thailand. During March, the tsunami energy was focused to the southwest, away from any nearby land mass, except Sumatra island which itself blocked significant wave activity towards Thailand and Malaysia (Eric Geist, 2005).
- The depth of focus of March 2005 earthquake and December 2004 earthquake were same 30 km below the sea bottom, but material at depth might be different, i.e. magma and solid rock respectively. Manimaran (2007) pointed out that the crustal break in Indo-Australian plate formed due to oceanic lithospheric shortening might have initiated the upwelling of magma along this break to the higher crustal levels. Moreover, the epicenter of March earthquake was located at near east of Indo-Australian plate crustal break. Hence it is more probable to locate the focus of 30 km depth in a upwelled magmatic zone. The partial energy of the earthquake might have been transferred to the magmatic zone, which inturn might have reduced the risk of tsunami during March 2005.

- In February 2005, a Royal Navy ship detected large landslides over the rupture zone of 2004 earthquake. Large thrust ridges (1500m high), an Oceanic trench several kilometres wide and displacement of huge water column were also recorded. However, no similar developments were detected associated with the 2005 earthquake which could also explain the absence of strong tsunami waves during March 2005 earthquake (Rastogi, 2005).

In addition to these differences, few other earthquake events in this region as described in the following paragraphs suggest differences in generation of weak and strong tsunami events due to varied parameters.

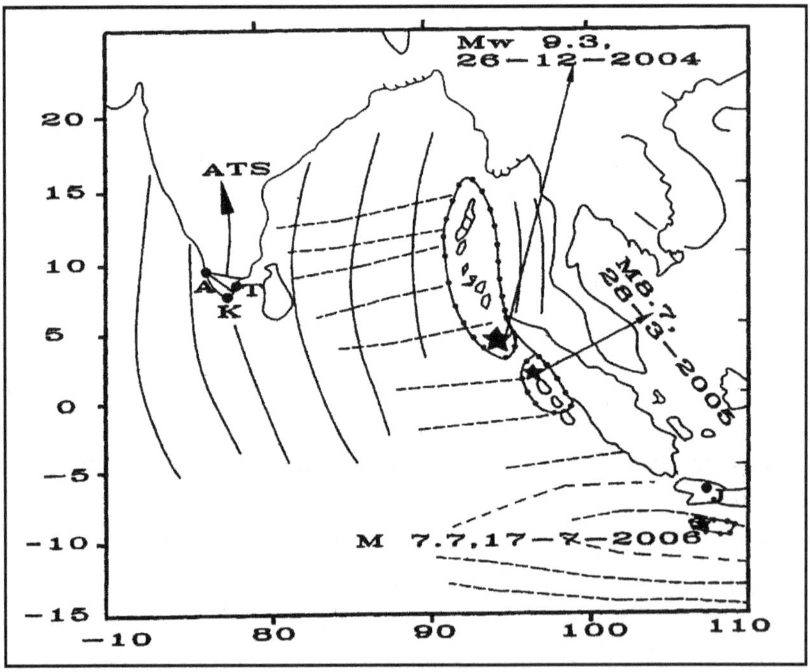

Figure 3. Map showing directions of movement of strong and weak tsunamis. The figure shows northerly moving crescent shaped zone of after shocks related to 26th December 2004 earthquakes **(Box 1)** and Southeasterly moving zone of aftershocks related to 28 March 2005 **(Box 2)**. The parallel Lines pointing the East and West focusing directions of 26 December Tsunami and southwest focusing of 28 March, 2005 Tsunami. Epicenter of 24 July 2005 (M 7.3) strong Andaman earthquake is also shown **(Location box 3)**, North and South Focusing of 17th July 2006 **(Box 4)**, Java Tsunami. ATS- Achankovil-Thamiraparani Shearzone. A-Allepy, T-Thiruchendur, *K-Kanyakumari*.

An earthquake of M 7.3 on 24th July 2005 at 9.12 P.M. (7.901° N and 92.131° E), 135 km west of Misha, Nicobar Islands was recorded. It had created a new rupture of 15 km by 80 km patch along a fault located on the western boundary of the crescent - shaped aftershocks zone of December 26th 2004 Sumatra earthquake (Fig.3). During this event, shocks were felt at south India (Chennai, Srikakulam and Vishakapattinam), Dum Dum - West Bengal, Phuket- Thailand, Kandy- Sri Lanka and Male - Maldives (USGS, 2005). Seiches were recorded at coastal belt of Kanyakumari district, while sea receded up to 50-70 meters at Tuticorin, south India. No significant tsunami event occurred following this earthquake due to strike - slip faulting and negligible vertical slip along the rupture.

A tsunamigenic major earthquake of magnitude 7.7 occurred on 17th July 2006, at 355 km south of Jakarta, Indonesia between Australian Plate and Sunda Plate. It occurred at shallow depth of 34 km and about 50 km north of Java trench where Australian plate thrusts beneath the Sunda plate and moves NNE at about 60 mm/ year (Lasitha et al., 2006). Tsunami of two meter high hit the coastal towns of Pangandaran, Serang, Yogyakarta and Denbasar following this earthquake. Hotels, buildings and paddy fields were worst affected at Pangandaran tourist city and more than 600 people died. A tsunami wave height of 60 cm reported at Christmas Island located at 225 km SW of the epicenter.

All these events and explanations listed above indicate, although submarine earthquakes, volcanism and landslides trigger tsunamis, the impacts vary significantly from weak to severe depending on complex parameters unique to each earthquake.

TSUNAMI RISK REDUCTION AND EARLY WARNING

As tsunami waves follow the dynamics of shallow water, the ocean floor topography or bathymetry governs the direction of propagation. Tsunami wave energy extends from the surface to the bottom even in the deepest waters. Wave amplitudes are decreased by the presence of coral reefs and mangrove forests bordering the coast. Waves can bend around circular Islands without great loss of energy, but become considerably smaller on the backsides of elongated angular Islands and continents. This is how the dominant control of bottom topography over propagation of tsunami waves and dissipation of tsunami wave energy could be explained. This nature also explains the observation that is why specific coasts are battered by tsunamis while adjoining regions are not affected (Devid Stevenson, 2005). Based on these characteristics, following strategies are normally followed to reduce tsunami risk.

- Avoiding a tsunami hazard area for any developmental work.
- Strict implementation of Coastal Regulation Zone rules and relocation of makeshift settlements to safer regions.
- Slowing strategy to reduce the destructive power of tsunami through friction by establishing strategically aligned forests, ditches, slopes and berms.
- Development of natural vegetation barriers along coastlines such as mangrove forests.
- Construction of steering walls and ditches to guide the force of tsunami away from important coastal structures and towns and blocking the force of tsunami through hardened structures like seawalls, compacted terraces and parking structures (Fig.4).
- Alteration of coastal bottom topography through positioning and constructing artificial and natural barriers or combination of both in such a way to dissipate or reduce the impending large waves including tsunami and cyclonic waves.

A rudimentary tsunami warning system was established in 1948 at Hawaiian Islands to provide warning information to the civil and military authorities around Pacific. However, 1960 Chilean and 1964 Alaskan tsunamis emphasized the need for an International Tsunami warning System (ITWS). The Pacific Tsunami Warning Centre (PTWC) in Honolulu Ewa Beach, Hawaii, established in 1949, serves as the regional Tsunami Warning Centre for Hawaii and as national/ international warning centre for tsunami that pose a Pacific-wide threat. This network comprises hundreds of seismic monitoring stations world wide as well as coastal tide gauges, besides sophisticated Deep-Ocean Assessment and Reporting of Tsunamis (DART) buoys capable of detecting even a few centimeter difference of sealevel. Many other regional tsunami-warning systems have been established in the Soviet Union, Japan, Alaska and Hawaii.

Functioning of these warning systems begins with the detection by any participating seismic observatory of an earthquake of sufficient size to trigger the alarms, set at the threshold of 6.5 on the Richter scale. PTWC collects the seismic data, locates the earthquake and computes its magnitude. When reports from tide stations show that a tsunami has been generated which poses a threat to the population in part or the entire Pacific, a warning is transmitted to the dissemination agencies for relaying to the public. The agencies then implement predetermined plans to evacuate people from endangered area.

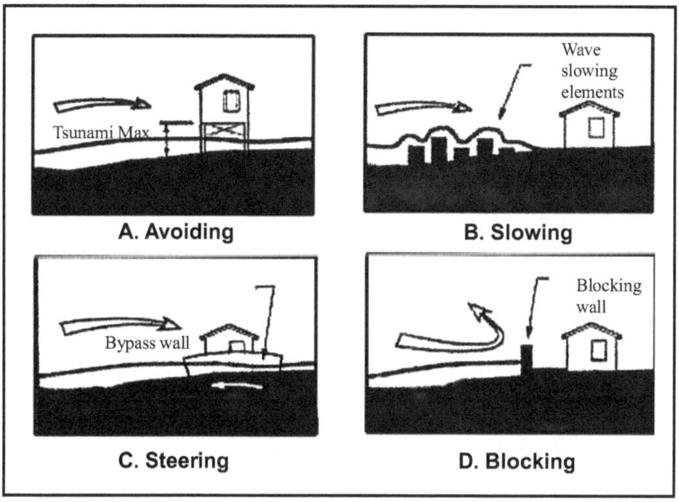

Figure 4. Model structures to reduce Tsunami risk

After December 2004 Indian Ocean Tsunami, the Indian Ocean Tsunami early warning system was established at the cost of 30 million US dollors, which incorporates the needs of storm surge forecast too (Gupta, 2005). Information from this warning system would be shared through the Indian Ocean - Global Ocean Observation System (IO-GOOS) partner countries.

CONCLUSIONS

This chapter has demonstrated that not only the magnitude of submarine earthquake, but also the areal extent of rupturing, time duration of rupturing, associated vertical uplift of plates, the material (magma or rock) at the depth of focus of earthquake, quantum of energy transferred to water column, direction of propagation of tsunami are the factors, in addition to many other parameters such as coastal topography and alighment of landmass with reference to probagating tsunami waves that control the impact of tsunami. A review of geotectonic characteristics, history of tsunamis all over the

World have indicated that each tsunami event is unique and hence evolving a generalized model for tsunamigenic earthquake/landslide/ volcanic eruption would be a far more complex geological problem that needs to be addressed immediately. Under this scenario, in order to provide effective tsunami warning, development of precise models of tsunami propagation under different rupture conditions for same tsunami prone area is required.

References

Abe, K., 2005 The Indian Ocean Tsunami (www.eri.u-tokyo.ac.jp/SUMATRA2004 /abe.html).

Appadurai Pillai, K., 1941 Kumari Kandam, The South India Saiva Siddhanta Works Publishing Society, Tinnevelly. 96p.

Bilham, R, Engdahl, E.R., Feldl, N and Satyabala, S.P., 2005 Partial and complete rupture of Indo-Andaman Plate boundary 1847-2004. Seis.Res.Lett. v.18. pp. 1-21.

EERI Survey., 2005 December 2004 Indian Ocean Tsunami. EERI-Indian Survey results. www.eeri.htm.

Geist, E., 2005 USGS Research Sheds lighs on why the March tsunami was smaller than the December tsunami. USGS, URL:http://www.usgs.gov/ newsroom/article.asp? ID=697.

Gahalaut, V.K, and Chanda. R.K., 2005 Sumatra Tsunamigenic Earth quake of 26 December 2004. DCS-DST NEWS, Govt. of India, January 2005. pp.18-20.

Geological Survey of India, 2005 Special issue on Tsunami, May 2005. Geological Survey of India Southern Region, Hyderabad. 52p.

Gorden, R.G. Demets, C and Argus, D.F., 1990 Kinematic constraints on distributed deformation in the Equatorial Indian Ocean from present motion between the Australian and Indian Plates. Tectonics. v.9. pp.409-422.

Gupta, H.K., 2005 Early Warning System for Oceanogenic Disasters in Indian Ocean (Tsunami and Storm Surges): The Indian Initiative. Jour.Geol.Soc.India. v.65. pp.639-646.

Harinarayana, T, and Naoshi Hirata., 2005 Destructive earthquake disastrous in the Indian Ocean: What next? Gondwana Res. v.8. pp.246-257.

Iyengar, R.N. and Radhakarishna, B.P., 2005 Evolution of the western coastline of India and the probable Location Dwaraka of Krsna: Geological perspectives. Jour.Geol.Soc.India. v.66. pp.285-292.

Lasitha, S., Radhakrishna, M and Sanu, T.D., 2006 Seismically active deformation in the sumatra-Java trench - arc region: Geodynamic implications. Curr.Sci. v.90. pp.690-696.

Manimaran, G. and Roychacko, P.T., 2006 Indian Ocean Tsunami - A Historical Perspective. Jour.Geol.Soc.India. v.8. pp.163-166.

Manimaran, G. and Ravindran, A., 2007 A caution on forthcoming volcanogenic tsunami in Indian Ocean. Res.Jour.Chem.Env. v.11. pp.19-22.

Manimaran, G., 2007 Indian Ocean tsunami and related events. Renuga Publications, Tirunelveli. 72p

Manimaran, G., and Renuga, K., 2007a Understand and conquer the seismic earth (Tamil) Renuga Publications, Tirunelveli. 250p.

Manimaran, G and Renuga, K., 2007b Tsunami: A competent scientific analysis (Tamil). Renuga Publications, Tirunelveli. 200p.

Narayana, A.C., Tatavarti, R. and Shakdwipe, M., 2005 Tsunami of 26 December 2004: Observations on Kerala Coast. Jour.Geol.Soc.India. v.65. pp.239-246.

Ramanujam, N., Nathakirimurugan, K., Ravindran, A., Arul Alphonse, P. and Sudarsan, R., 2005 Impacts of December 26th 2004 tsunami over the Gulf of Mannar and coastal aquifer system in southern part of Tamilnadu, India. In: Ramasamy, SM. and Kumanan, C.J. (Eds.), Tsunami : The Indian Context, Allied publishers. pp.231-238.

Ramasamy, S.M. and Kumanan, C.J., 2005 (Editors) Tsunami : The Indian Context, Allied Publishers. 315p.

Rao, P.N. and Hanmanthachary, A., 2005 What caused the great Sumatran earthquakes of 26 December 2004, and 28 March 2005? Curr. Sci. v.89. pp.449-452.

Rastogi, B.K., 2005 Why did the 28 March, 2005, Sumatra earthquake of Mw 8. generate only a minor tsunami? Curr.Sci. v.89. pp.731-732.

Sivathamby, K., 1986 Literary History in Tamil. Tamil University, Thanjavur. 200p.

Stein, S., and Okal, E.A., 1978 Seimicity and the tectonics of the Ninety east ridge area: Evidence for internal deformation of the Indian Plate. Jour.Geophys.Res. v.83. pp.2233-2248.

Stein, S. and Okal, E.A., 2005 http://www.earth.northwestern.edu/people/seth.

Stevenson, D., 2005 Tsunamis and Earthquakes: What physics is interesting? Jour.Geol.Soc. India. v.66. pp.497-499.

USGS., 2005a earthquake hazards program. Largest earthquakes in the world since 1900, EQ Facts & Lists, Earthquake Lists. (http://neic.usgs.gov/neis/eqlists/10maps-world.html).

USGS., 2005b Magnitude 9.0 Sumatra-Andaman islands earthquake, off the west coast of northern Sumatra, 2004 December 26 at 00:58:53 UTC, (http://earthquake.usgs.gov/ eqinthenews/2004usslav/).

USGS., 2005c Magnitude 8.7 - Northern Sumatra, Indonesia 2005 March 28 16:09:36 UTC http://earthquake.usgs.gov/eqinthenews/2005/usweax.

USGS., 2005d A major earthquake magnitute 7.3 - Nicobar Islands, India region 2005., July 24, 15:42:06 UTC, Earthquake Hazards program. http://earthquake.usgs.gov/ recenteqsww /maps/10/90-10.html.

Wikipedia., 2005a Sumatra earthquakes occured at 16.09:37 UTC on March 28. 2005. http://en.wikiepedia. org/Wiki/2005-smatra-earthquake, 8, May 2005.

Wikipedia., 2005b http://en.wikepedia.org/wiki/2004. Indian Ocean earthquake of 1 01k.9Jul. 2005.

Yagi., 2005 Preliminary Results of Rupture process for the march 28, 2005, Northern Sumatra, Indonesia earthquake. yagi@IISEE, BRI 2005.

Yamanaka, Y. 2004 04/12/26 OFF West Coast of Northern Sumatra Coast of Northern Sumatra, (http://www.eri.u.tokyo.ac.jp/sanchu/seis.note/2004/EICI61 e.html).

❑❑❑

4

Andaman-Nicobar Island Arc in the Evolving Tectonic Scenario of Bay of Bengal

***BARENDRA PURKAIT & **P.K.GANGOPADHYAYA**

* Geological Survey of India, Environmental Geology Division, Salt Lake, Kolkata - 700 091. E-mail: <baren_purkait@yahoo.co.in>

** Department of Geology, Presidency College, Kolkata – 700 073.

Detailed information on tectonic evolutionary history of the Bay of Bengal is presented followed by elucidation of these characteristics towards tsunamigenesis in this chapter. The Andaman-Nicobar Island Arc (ANIA) is situated on a 1000 km long plate margin and represents a key region in the developing oceanic crustal area of Bay of Bengal. Frequent earthquakes including the tsunamigenic earthquake of 26th December 2004 off the coast of Sunda take place in this region. The subduction of Indian plate under Myanmar-Thailand plate drives formation of chain of Andaman-Nicobar group of Islands. These oceanic crust formation, subduction and arc formation are all seismigenic and generate shallow focus earthquakes in the Bay of Bengal. It is also observed that the seismicity is decreasing as the bathymetry increases towards SW of the Bay of Bengal. Thus, in long term, the slow accumulation of stress as tectonic plates converge may result in many more earthquakes in this region. After a dormancy of over 100 years, the Barren Island volcano renewed its activity following recent earthquakes. The seismic activity along the Andaman-Nicobar trench is continuing with earthquakes of magnitude varying from 4.8 to 7.2 Richter scale. All these observations suggest resurgent and intense tectonic and volcanic activities due to which allied hazards such as tsunami may take place in future.

INTRODUCTION

The evolving Earth, depending on its heat flow pattern in the mantle throughout geological time, had passed through stages, which are diagnosed through characteristic geodynamic signatures. This chapter focuses on a very active region in the crust, namely, the Andaman-Nicobar Island Arc (ANIA) to recognize such signatures and decipher relative safety of surrounding regions including east coast of India. As similar active arc systems such as Sunda Arc & Japan Arc generate volcanic, seismic and tsunami hazards, understanding the ANIA would be of help assess safety of surrounding regions. Abundant geoscientific literature exists on the northern part of the Indian plate, which is undergoing compressive stress since collision between the Indian and Eurasian plates about 55-45my. However, the behavior of remaining part of the Indian plate underneath the Indian Ocean is still sparcely studied. Owing to these reasons, this chapter attempts to fill the gap in our understanding as well as to assess the implications of renewed phases of geohazard occurrences including that of catastrophic earthquake and tsunami of 26th December 2004.

CRUSTAL CONFIGURATION OF BAY OF BENGAL

The approximate pattern of flight of the Indian plate since the disintegration of the Gondwanaland around 180 million years ago had been outlined by Le Fort (1975) in relation to opening phases of the Indian Ocean (Table 1). The flight patterns of the continental shield of India is correlated to magnetic anomalies in the Indian Ocean in a time clock as follows - anomaly No. 32 (71 my.), anomaly No. 24 (55 my.), anomaly No. 13 (38 my.) and anomaly No. 5 (10 my.). The plate motion suggests that the continental shield had paused twice and rotated anticlockwise during its northward flight coeval with opening up of Indian Ocean in phases. The tectonic arrangement for the Indian and Australian plates and intervening oceanic parts had been shown by Condie (2005), the excerpts of which are as presented herein.

The eastern part of the oceanic extension of the 'Indian plate' defines a southwestwardly convex arc. This oceanic sector of the 'Indian plate' is subducting eastwards below an agglomerated complex of Myanmar sub-plate, Sunda-Sumatra (S-S) Arc-Trench System and isolated continental blocks of Malaysia, Thailand and Philippines.

A remarkable feature is that the nearly N-S trending Japan arc system, bordering the eastern margin of the Philippine plate joins the easterly trending boundary of Indian plate (marked by the S-S arc-trench system) near Borneo almost at right angles, forming a T-shaped junction.

Table 1. Stages of Indian plate movement and anomalies in Indian Ocean (*after* Le Fort, *1975)*

Anomaly no.	Age (10^6 year)	Interpreted Event
1	0	Spreading rate: 4 to 5 cm/year
3	4-5	(Possible readjustment)
5	9-10	Movement resumes
6	20-21	Slow down
12	35	Movement resume towards N35^{0}E
18	45	-
21	55	Significant slow down (stop?)
		Rapid movement of India, slightly E of N
	75	Halt spreading rate up to 8 cm/year
		Oldest recognized magnetic anomaly
32	100	Gondwana in four blocks
	140	Gondwana in two blocks
	180	Beginning of Gondwana partition

Considered in detail, there is difference between the West Indian Ocean (WIO) crust and the East Indian Ocean (EIO) crust as reflected in bathymetry, earthquake epicenters, identified magnetic anomalies and fracture zones (Fig. 1). In the EIO, the long Ninety-east Ridge could be a piece of uplifted oceanic (Laughton et al., 1972) ridge. It is currently speculated that the ridge is a signature of hot-spot track (Condie, 2005). The possible nature of the crust in Bay of Bengal and adjoining Indian Ocean region, deduced from geophysical lineament studies, appears to be rather complex (Singh, 1989). In this study an attempt had been made to develop a method for isolation of crustal contribution in the MAGSAT data and to prepare a map showing the anomalies over the Indian continental and oceanic regions (Fig. 2). The occurrences of positive and negative magnetizations across the Indian shield and only the positive magnetization in the Bay of Bengal indicate that continental crust seems to continue into the Bay. On the other hand, the volcanic Andaman-Nicobar Arc and Sunda-Sumatra arc are active signatures of oceanic crust under development. The eastern part of Indian plate, therefore, shows variation from abyssal depth to rises (Ninetyeast Ridge) and volcanic arc-trench systems. Ongoing oblique subduction of this sector has attracted much attention after the catastrophic earthquake and tsunami on 26th December 2004.

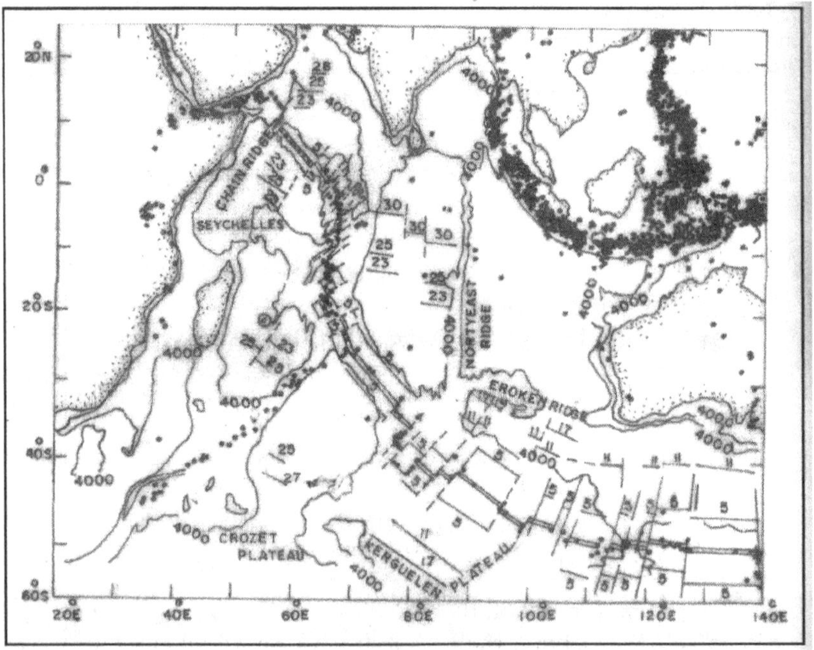

Figure 1. Summary of Indian Ocean bathymetry (1 and 4 km contours), earthquake epicenters, identified magnetic anomalies and fracture zones (*after* Laughton et al. *1972*).

SEISMICITY AND TECTONICS OF BAY OF BENGAL

Geological Survey of India had published a collection of papers under the title "Geophysical exploration basement type and seafloor spreading" in Special Publication No. 29 in 1992, which forms the most recent and comprehensive geophysical observations on Bay of Bengal and the Andaman Sea. Few relevant observations pertaining to the objectives of present chapter are listed herein.

- Based on recognition of presence of distinctly different power spectra and magnetic anamolies of Andaman Sea from that of other sectors in and around Andaman sea, Sengupta et al. (*op.cit.*) concluded that the Andaman sea is a composite mosaic of oceanic and continental types, although its precise configuration could not be deduced.

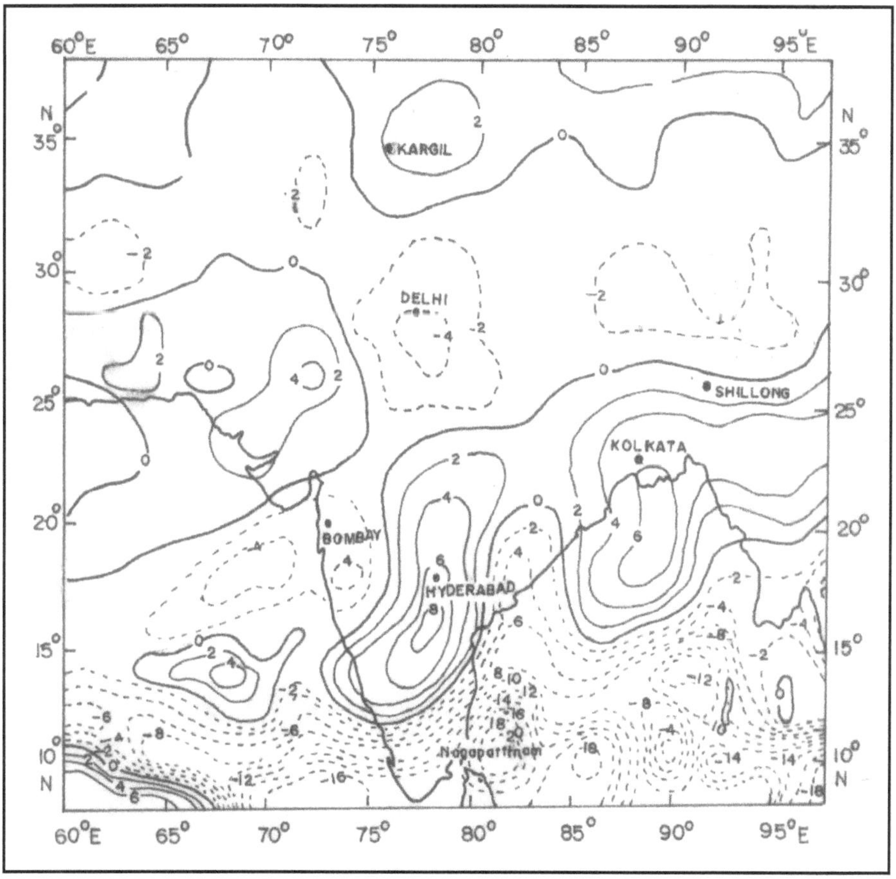

Figure 2. Map showing magnetic moment distribution calculated for $2^0 x 2^0$ blocks for India and contiguous regions by inverting anomalies in X and Z jointly (contour interval $2x10^{13} Am^2$) *(after* Singh, *1989)*

- Venkatateswarlu et al. (*op.cit*) made observations on marine magnetic anomalies and prepared a map of basement structures for part of Bay of Bengal (Fig. 3). A prominent magnetic anomaly occurring as two separate closures indicates the presence of submerged volcanoes off Machilipatanam. This location of the volcano falls in the northeast extension of a deep main fault in the Peninsular India. The magnetic anomalies have delineated the extensions of 85^0 E ridges to the north and its arcuate bend towards northeast.

- Mukhopadhyay et al. (*op.cit*) have recorded geophysical evidence for crustal transition under Eastern Continental Margin of India (ECMI) by preparing a revised Free-Air Gravity anomaly map. Gravity modeling suggests a transitional crust below the ECMI whose thickness varies from 16 to 24 km. The models postulate a thicker oceanic crustal wedge seaward of the transitional crust in most part of the ECMI.

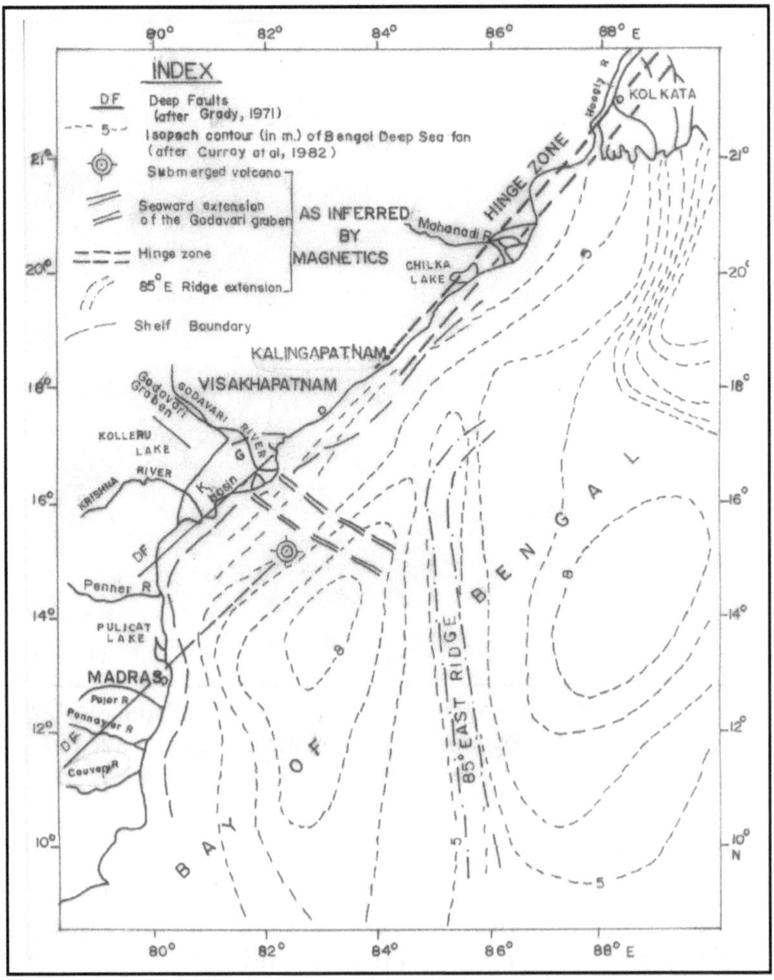

Figure 3. Basement structures as inferred from the magnetic data collected from RV Samudra Manthan during 1983-88 and from published sources along east Coast of India (*after* Venkatateswarlu, 1992).

Hence, the ECMI is interpreted as a typical Atlantic-type margin that evolved following the Gondwanaland breakup by Late Cretaceous and northward drift of Indian plate. In the light of these observations, the nature of seismic activity and tectonics of the region has been interpreted as detailed herein.

The devastating earthquake of 26th December 2004 has aroused renewed interest in plate movements in relation to seismicity throughout northeastern Indian Ocean and Bay of Bengal region in particular. Two sets of USGS Website releases give a detailed picture of seismicity (1900-2004, M=6) and inferred tectonic setting in Northeast Indian Ocean Region.

Biswas and Majumder (1996) have studied seismicity and associated fault planes of several earthquakes in the Bay of Bengal. They have documented a rather dispersed diffuse seismicity and thought that this phenomenon is due the fact that the intraplate region of the entire Bay of Bengal and Peninsular India are seismically active of which the Gangetic delta and the eastern flank of the Ninetyeast Ridge show relatively higher seismicity in comparison to the thickly populated epicenters at the eastern plate boundary (Fig. 4). Predominance of thrust faulting earthquake mechanisms over fault planes dipping north and approximately N-S pressure axes imply dominance of N-S compression in the intraplate region of the Bay of Bengal.

Seismic and tectonic setting of the northeastern part of Indian plate, particularly the subduction zone was studied by Dasgupta and Mukhopadhyay (1997). According to these authors, tectonic framework of the Barren island volcano is delineated by a set of seismically mapped faults, including the west Andaman fault, which is connected to the Semangko Fault in Sumatra. The morphology of the Andaman Benioff Zone as seen on a vertical section reveals a seismic gap at 90-110 km depth. Spatially it relates to the zone of current volcanism of the Barren Island and its neighboring Alcock seamount in a structurally depressed area. The seismic gap in the Benioff zone is interpreted as a zone of partial melting of descending Indian Ocean lithosphere, which prevents stress accumulation due to decreased viscosity. Pronounced back-arc seismicity at shallow depths is seen further to the east and southeast of Alcock seamount.

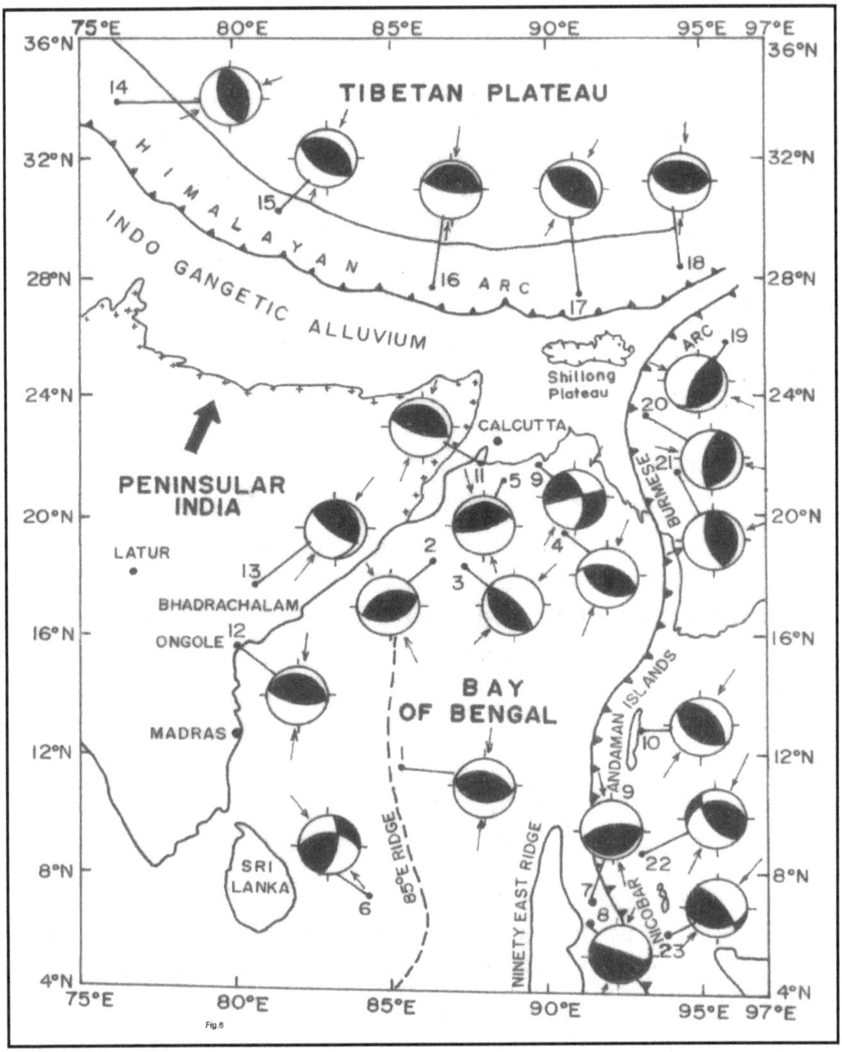

Figure 4. Fault plane solution of the Bay of Bengal earthquakes (1-10) with other plate boundary and coastal earthquakes (11-23, see Table 1) is shown on a tectonic map. Black and white region of each of each fault plane solution represents areas of compression and dilation. Fault plane solution with converging arrows indicates the direction of compression. Lines with solid triangle are the active thrust zones and mark the northern and northeastern plate boundaries. Broken line along 85^{0}E longitude represents the trend of 85^{0}E Ridge. Line with small plus symbol shows the schematic boundary between crystalline peninsular Indian shield from the Indo-Gangetic alluvium. Large arrow represents the direction of the Indian plate movement (*after* Biswas and Mazumder, *1996*).

The seismicity map indicates widely diffuse intraplate seismicity in the Bay of Bengal and adjoining Peninsular India in comparison to the clearly defined higher seismicity at the eastern Indian plate boundary. A relatively higher seismicity is noted at the Gangetic delta in the Bay of Bengal where a thick pile of sediment column occurs and active sedimentation takes place. Another cluster of epicenters is observed in the eastern flank of the Ninety East Ridge possibly evolved due to the interaction of ridge and the subduction zone. Distinctly N-S defined thickly populated epicenters situated further east of the Bay of Bengal make the eastern Indian plate boundary where Indian lithosphere subducts under the Burma-Andaman-Nicobar Island Arcs. In the Indian continental shield, a relatively higher seismicity is observed to the west of Kolkata, central eastern coastal and southern parts of Peninsular India. The whole Precambrian shield of Srilanka remains aseismic. In contrast, the entire sea floor of the Bay of Bengal is seismicprone implying ongoing intraplate deformation in the sea floor.

At the present state of knowledge, the geodynamic scenario can be briefly stated as follows. Below the Andaman arc, the Indian Ocean lithosphere actively subducts under the overriding Burma sub-plate (Fitch, 1970; Curray et al. 1979, 1982). The subduction azimuth varies from oblique in the Andaman and Sumatra region to frontal subduction south of Java (McCaffrey, 1994). The Neogene back-arc spreading activity below the Andaman Sea is inferred to be due to leaky transform tectonics (Eguchi et al. 1979).

VOLCANIC EVENTS IN ANDAMAN-NICOBAR ISLAND Arc

The Andaman Islands themselves are accretionary prisms with their top eroded. The sediments are scraped from the seabed due to the subduction of the oceanic basement. Due to the erosional activity, the older Cretaceous Baratang and Port Meadows formations are exposed on the western margin of islands and progressively younger sediments are exposed eastwards. Little Andaman (with coral top), Ritchies Archipelago and Nicobar Islands are all accretionary prisms of sediments, which appeared later. The subduction of the Indian plate takes place east of the island in Dilligent strait. From north to south, there is a row of folds that can be traced from west coral bank to Middle bank-South bank-North Sentinel Island-Sentinel Island and finally upto Darlymple bank (Avasthi, 2005).

The Island arc system of the Andaman is essentially built up around Narcondam and Barren islands. The subduction of Indian plate under Myanmer-Thailand plate causes the formation of the chain of Andaman and Nicobar group of islands. As a result, frequent earthquakes take place. The 26th December, 2004 earthquake off the coast of Sumatra, the resultant tsunami, thousands of aftershocks along the stretch running from Sumatra to the Andaman Islands including that in January, 2005 indicate resurgent seismicity in the region. The 26th December 2004 earthquake had probably triggered 28th March 2005 earthquake off Sumatra - the second largest in 40 years. The last large tsunami to strike Indonesia occurred in 1861. Thus, in long term, the slow accumulation of stress as tectonic plates converge makes earthquakes inevitable in this region.

The Barren Island volcanism during 1991 and 1994-95 has drawn much attention to resurgent thermal activity in the eastern periphery of the Indian plate (Halder and Luhr, 2003). The Barren Island volcano belongs to the SE Asian volcanic arc. Neogene-Quaternary and present day volcanism in this region is related to this Benioff zone. Barren Island volcano, rising 2300 m. above the Andaman sea floor, has displayed volcanism in three recognizable phases: the oldest cycle was during Late Pleistocene age, followed by the second cycle between 1787 and 1832 AD and the third cycle between March-October, 1997 and December, 1994-May 1995. The volcano is now in active fumarolic stage. The 1991 eruption started as Strombolian type and it later merged into Plinian type; the 1994 - 1995 eruption was of Strombolian type. The Pleistocene lava is olivine basalt; the second cycle is olivine basalt but impoverished in MgO. The 1991 lava is olivine bearing basaltic andesite while the lava of 1994 - 1995 is again olivine basalt similar to that of the second cycle. Lava erupted in all these three cycles was tholeiite being composed of a melt and silicate megacrysts of olivine (F_0 87-70) with included chrome spinel (Cr_2O_3 - 8%), clinopyroxene ($Mg_{48\text{-}44}Fe_{16\text{-}10}$) and plagioclase ($An_{95\text{-}55}$) as reported by Halder and Luhr (2003). Lava chemistry closely compares with that of Island Arc tholeiite enriched in LREE and LILE and depleted in HFSE composed to N-type MORB. A comparative study of major and minor element chemistry of Barren Lava with that of N-type MORB, OIT, IAT and Sunda Arc Basalt indicated that Barren lava compares well with that of IAT and the Sunda Arc basalt. It is concluded that the Barren Lava (BL) are IAT in composition and possibly generated through partial melting of an e-type mantle source (Halder and Luhr, 2003). This specified partial melting zone possibly occurs at a depth of about 100 km below the sea floor, which is the zone of seismic gap in the Andaman Benioff Zone below the BV. It may be noted that the average heat flow below the Barren volcano is abnormally high (nearly 10-12 times greater than that in the adjacent areas) indicative of sustained convection from mantle source.

CONCLUSIONS

The Andaman-Nicobar Island Arc is situated on a nearly 1000km long active plate margin. As a result, frequent earthquakes take place. Geotectonic model suggests oblique convergence between the India-Australia plates and the India-Myanmar-Andaman-Sumatra blocks. This has resulted in Andaman-Java subduction zone and an easterly dipping Benioff Zone. Neogene-Quaternary and the current volcanism in this region are found to be related to this easterly dipping Benioff Zone. The subduction of the Indian plate takes place east of the islands in Dilligent Strait. The subduction of Indian plates under Myanmar-Thailand plate causes the formation of the chain of Andaman-Nicobar group of Islands. Review on geodynamic signatures, ongoing seismic and volcanic activities of the region and composition of lava flows indicate active tectonic, volcanic and tsunamigenic processes in the Bay of Bengal and adjoining Indian Ocean.

Implication of these interpretations towards safety of the countries encircling Bay of Bengal is that, as the region is undergoing a renewed phase of activism, proper remedial measures and contingent plans for distaster preparedness have to be undertaken immediately. Although the geological processes as detailed in this chapter go on for millions of years albeit at different paces, with the increased utilization of vulnerable locations for domestic, industrial, commercial and recreational purposes, the vulnerability levels were stacked against the safety of human lives and infrastructure. Hence, appropriate measures are necessary.

References

Avasthi, D. N., 2005 The Sunda trench and the Andaman-Nicobar Islands. Jour.Geol.Soc.India. v.66. pp.519-520.

Biswas, S. and Mazumdar, R.K., 1996 Seismicity and tectonics of the Bay of Bengal: Evidence for intraplate deformation of the Northern Indian Plate. Tectonophys. v.457. pp.1-13.

Condie, K.C., 2005 Earth as an Evolving Planetary System. Elsevier, Amsterdam. 447p.

Curray, J.R., Moore, D.G., Lawver, L.A., Emmel, F.J., Raitt, R.W., Henry, M. and Kieckhefer, R., 1979 Tectonics of the Andaman Sea and Burma. Amer.Assoc.Petrol.Geol.Mem. v.29. pp.189-198.

Curray, J.R., Emmel, F.J., Moore, D.G., and Raitt, R.W., 1982 Structure, tectonics and geological history of the northeastern Indian Ocean. *In:* Nairn, A.E.M. and Stehli, F.G., (Eds.). The Ocean Basins and Margins, Plenum.Publ.Corp. v.6. pp.399-540.

Dasgupta, S., and Mukhopadhyay, M., 1997 Seismicity of the Andaman subduction zone and recent volcanism. Jour.Geol.Soc.Ind. v.49. pp.513-521.

Eguchi, T., Uyeda, S. and Maki, T., 1979 Seismotectonics and tectonic history of the Andaman Sea. Tectonophys. v.57. pp.35-51.

Fitch, T.J., 1970 Earthquake mechanisms in the Himalayas, Burmese and Andaman regions and continental tectonics in central Asia. Jour.Geophy.Res. v.75. pp.2699-2709.

Geological Survey of India., 1992 Recent Geoscientific Studies in the Bay of Bengal and the Andaman Sea. Special Publication No. 29. 278 p.

Haldar, D., and Luhr, J.F., 2003 The Barren Island Volcanism during 1991 and 1994-95: Eruptive Style and Lava Petrology. Mem.Geol.Soc.India. v.52. pp.313-338.

Laughton, A.S., Mckenzie, U.K. and Selater, J.G., 1972 The structure and evolution of the Indian Ocean. Proc.24th Internat.Geol.Cong. Montreal, Section-8. pp.65-73.

Le Fort, P., 1975 Himalayas: The Collidel Range. Amer.Jour.Sci. v.275-A. pp.1-44.

McCuffrey, R., 1994. Global variability in subduction thrust zone - Forearc systems. PAGEOPH. v.142. pp.173-224.

Singh, B.P., 1989 MAGSAT in lineament studies: Results from Indian region. Mem.Geol.Soc. India. v.12. pp.181-188.

❑❑❑

5

Assesement of Tsunami Hazard along Tangapatnam - Ovari Coast, Tamilnadu, using Remote Sensing and GIS Techniques

M.RAJAMANICKAM & S.RAJENDRAN
Department of Earth Sciences, Annamalai University, Chidambaram - 608 002.
E-mail: <sankaranrajendran@yahoo.com>

Need for systematic mapping and hazard zonation is emphasized in this chapter. Though prevention of natural disaster is not possible, its destruction could be minimized through implementation of effective hazard management system. With this motive, the coastal region between Tangapatnam and Ovari, Tamilnadu State has been classified into various categories of vulnerability based on its response to the 26th December 2004 tsunami. Based on the extent of seawater inundation and coastal geomorphic features, we have assessed the tsunami impact and estimated the behavior of the beaches in case of similar havoc in future. This study had shown that the maximum seawater inundation recorded in the study area is 750m in Colachel and the minimum is 100m in Kadiapatanam, Mandakadu and Vaniakudy areas. Chinnamuttom, Kanyakumari, Manakudy, Pallam and Colachel areas are under high risk.

INTRODUCTION

On 26th December 2004, tectonic disturbances happened near Sumatra Island with an intensity of 9.3 in the Richter scale wrecked havoc on southwest part of India. In addition to inflicting life and infrastructure loss, the resultant tsunami event had also eroded severely the coastal region and modified topography of beach and shallow coastal regions. It did raise the concern of scientists and emergency planners about the impact of larger earthquake and consequent tsunami should there be any recurrence of such event. Despite becoming aware of tsunami hazard, there has been confusion about areas at risk and areas of safety. This chapter is an attempt to assess the vulnerability of part of Tamilnadu coast based on its geomorphological conditions and the impact of recent tsunami. The hazard maps have been generated and are intended for educational purposes, to improve awareness of tsunami hazards and to encourage responsible emergency planning efforts by illustrating the range of possible tsunami events.

STUDY AREA

The study area (Fig.1) lies between N 8^{0} 04′ to N 8^{0} 17′ latitude and E 77^{0} 32′ to 77^{0} 54′ E langitude at southern and western part of the Tamilnadu. It encompasses the Kanyakumari and Tirunelveli districts. The study area is bounded by Indian Ocean in the south, Arabian Sea in the west and Bay of Bengal in the east but the main part of the coast faces the Arabian Sea with mountains and undulating valleys in the north. The study area is manifested with marine terraces, sand dunes, beach ridges, estuaries, floodplains, beaches, mangroves, peneplains, uplands, sea cliff, etc. Apart from the perennial river Thamirabharani, streams like Nambiyar, Hanuman, Palaiyar, Panniyar and Valliyar form the major drainage system of the study area with several other creeks and tidal channels. Almost the entire coastline is erosional in nature showing steep gradient at the beachface. Coconut plantation encircles most the beaches beyond the dune. Rich growth of mangrove and salt marshes has been observed in the beaches near estuary in Manakudy. Continental shelf along the study area is very wide.

METHODS AND MATERIALS

As tsunami hazard maps are essential tools for land use planning in tsunami-prone areas (Raval, 2005) present study had focused on mapping extent of on tsunami inundation, documentation of flood return periods for different areas, demarcating zones or areas likely to be affected by tsunami of different magnitudes and/or frequencies and probability levels, based on which suggested types of permissible developments in these zones, so that impact of future tsunami should one occur, can be minimized, if not avoided. A tsunami damage map is considered as a preliminary, yet necessary initial step for all regional development policies. Field investigations were supplemented by remotely sensed data in terms of high-resolution satellite data to estimate inundation area. In addition, close contour information, depth of inundation were also estimated and used in this study.

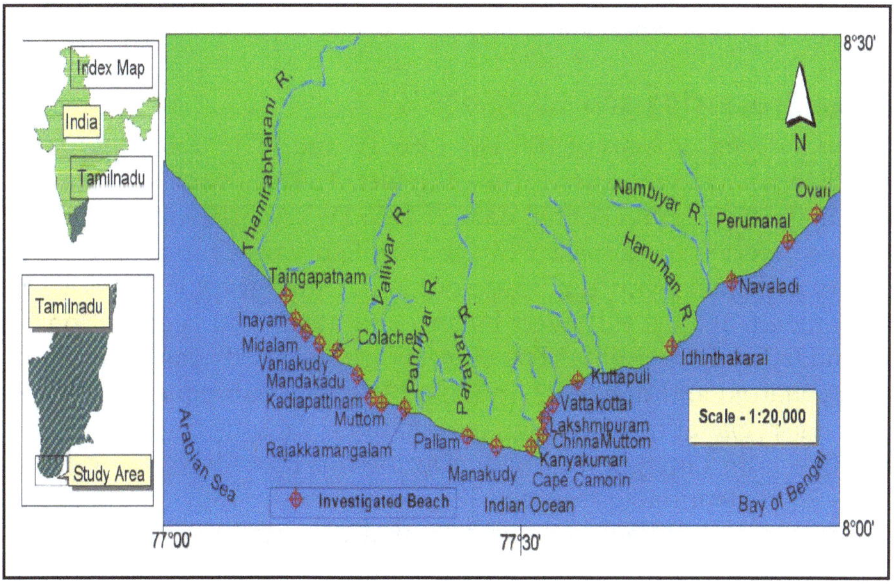

Figure 1. Location of the study area

Field Survey and Position Fixing

44.34 km long coastline between Thangapatnam and Ovari was mapped during this present study. Following the suggestions of Emery (1961), during the fieldwork, primary and secondary control points were identified and were utilized for the present study. Data on tsunami inundation and run-up levels were gathered from field examination of shoreline morphology and the extent of destructions (Photograph Nos. 1, 2, 3). Similar to the method followed by Jaffe and Gelfenbaum (2002), true wave heights of tsunami were measured and/ or estimated from the accounts of the interviewees (i.e. relative to their body or height of watermarks) and landmarks such as trees, rocks, coral reefs, dikes, riverbanks and other natural features found in the area. The run-up measurements were made along transects across coastline. Transects were made from lowtide water line upto maximum inundation point. The spacing between the two transects is about 2km as followed by Shuto (2001). The maximum inundation points were identified based on local enquiry and signatures such as degraded vegetation, deposits of floating garbage and sediment characteristics. The maximum elevation information was collected from hightide line to landward up to 1.5km. Beach profile survey through Stack and Horizon Method (La Fond and Prasada Rao, 1954) simplified by Emery (1961) and Koji Minoura (1994) was conducted using levelling and surveying equipments.

Image Processing and GIS Analysis

The digital images of IRS-P6 - LISS IV and PAN data and Quick Bird images were processed using the ERDAS Imagine 8.3.1 software. For digital image processing, data of LISS IV PAN (Resolution: 5.8m) was georeferenced as master image in which various ground control points collected from the Survey of India topographic maps were recognized. It was followed by incorporation and recognition of secondary control points. The projection type used was 'polyconic' with spheroid and datum as 'modified Everest'. Subsequently image-to-image registration of PAN image into LISS IV multi-temporal images and Quick Bird images was made. During this procedure, many sharp and easily identifiable features like road-crossing, railway crossings, canals, temple, bridges etc., were located on both images as ground control points. Additionally, control points such as bi-junction/tri-junction points from the cadastral map of Department of Survey and Land records were also incorporated.

Non-linear edge enhancement techniques were applied to delineate the land, water and vegetation layers from the LISS IV imagery. This enhancement technique was aimed at emphasizing edges in image data. During this operation, it delineated sharp edges of water and land boundaries. Non-linear edge enhancement was attempted using nonlinear combinations of pixels. The Roberts edge detector was also applied for delineation of water and land boundaries. It was based on 2 * 2 pixel window and the new pixel values were computed according to the equation. Resulted edge enhanced output layer has digitized land, water and vegetation layers. Finally, the delineated land, water and vegetation vectors were transferred into georeferenced cadastral map. Similarly, various thematic layers namely, land use/land cover, coastal geomorphology, shoreline change from 1969 to 2004 were prepared from Quick Bird and IRS P6 LISS IV images. In addition, the information on High Tide Line (HTL), Coastal Regulation Zone (CRZ) buffers, coastal census and land elevation contours were collected from secondary sources and were integrated into 1:5000 scale cadastral maps. The integration was performed in Arc GIS software version 9.1.

Photograph 1. In many of the regions of the studied coastal zones, thick tsunami deposits are common upto 500-750 meters inland from coastline.

Photograph 2. The Bridge connecting Manakudy and Keelamanakudy has been dismembered and is found parallel to the bank.

Photograph 3. Similar to Colachal, the coastal village Rajakkamangalam shows run up elevation of 10 meters above MSL.

The high tide line data obtained from the Department of Environment was plotted on the land use/ land cover map and buffers at a distance of 200m and 500m were generated to understand the inundation, land use/land cover pattern within the CRZ corridor. The buffer of 1.5 km was generated from Hightide line (HTL) and land elevation values obtained from theodolite and Global Positioning System were converted into shape file (.shp). After grigging of these values, 2 m contour interval maps were generated from HTL points towards land up to 1.5 km distance for 24 stations. The land elevation contours were converted into Triangulated Irregular Network (TIN) format in addition to generation of slope map. The cadastral TIN model was directly converted into three-dimensional surfaces, with which topographical changes with respect to inundation, land use/land cover and CRZ buffers of coastal stretch were analyzed.

The inundation distance was decoded by its signature in the coastal settings and from the local people's information. Digital Elevation Model (DEM) has been projected for the study area using Surfer package. Similar to the methodology followed by Walsh et al. (2000), tsunami hazard maps were prepared using Geographic Information System (GIS) technique based on the inundation distance with respect to the nature of the coast to show the inland extent of flooding and topography of the area.

RESULTS AND DISCUSSION

As the extent of inundation is controlled by the angle of incidence of the tsunami wave and its velocity, the very presence of Srilanka (Fig.2) between India and the epicenter of 26th December 2004 earthquake, most of the beaches along the east coast of India experienced only the 'shadow waves' (Keating et al. 2004; Besana et al. 2004; Minoura, 1997). However, the beaches along the west coast starting from Kanyakumari have experienced rapid movement of refracted waves. Based on this observation, together with the location of seismically active subduction zone of Bay of Bengal located adjoining Sumatra Island, we ascribe east coast of India as comparatively safe while the west coast is termed high-risk area owing to the absence of any natural barrier similar to that of Srilanka. It is inferred from table 1 that maximum inundation had occurred in the coast where there is a river mouth or an estuary as in the cases of Manakudy and Colachel. The inundation proved to be ineffective along the coast where rock exposures are present as in the cases of Muttom and Kadiapatanam (Chandrasekar and Loveson Immanuel, 2005). Though there are numerous river mouths in the east coast, inundation was less because of the fact that the approached waves were of low intensity due to the obstruction rendered by Sri Lanka. However, the thick natural and plantations of west coast have curtailed the incursion of tsunami waves.

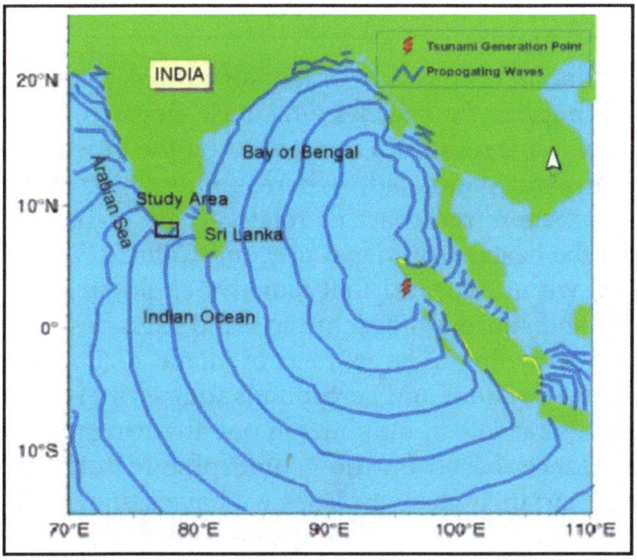

Figure 2. Nature and Angle of Incidence of the Tsunami waves

Table 1. Inundation Distance Extent along the Study area

Location	Longitude	Latitude	Elevation (m)	Inundation distance (m)
Ovari	77.49	8.17	19	150
Idinthakarai	77.45	8.14	18	175
Perumanal	77.39	8.09	17	200
Navaladi	77.37	8.08	16	200
Kuttapuli	77.36	8.08	16	250
Vattakottai	77.34	8.07	15	300
Lakshmipuram	77.34	8.07	16	250
Chinna muttam	77.34	8.06	17	350
Kanyakumari	77.33	8.04	21	300
Keelamanakudi	77.29	8.05	09	600
Pallam	77.25	8.05	14	400
Rajakkamangalam	77. 22	8.06	16	150
Muttom	77.19	8.07	11	200
Kadiapatanam	77.18	8.08	14	100
Mandakadu	77.16	8.09	16	100
Colachel	77.15	8.1	12	750
Vaniakudy	77.14	8.11	16	100
Midalam	77.12	8.12	17	300
Enayam	77.09	8.13	15	130
Taingapatnam	77.1	8.14	14	200

The inundation of seawater of tsunami waves could not proceed for longer distance in the beaches which are elevated, comparatively, from the mean sea level (MSL) as attested by the beaches of Muttom, Kadiapatanam and Mandakadu whereas inundation was maximum in the beaches where the coast is, relatively, at or near the MSL as evident from the beaches of Manakudy and Colachel (Fig. 3a and 3b). Nevertheless, we understand that tsunami could strike the coast at high velocity and cause infilling of bays and estuaries, which in turn could enforce drowning of large tracts of inland regions along coastal rivers. These observations and perceptions suggest that the geomorphic features play a vital role in minimizing or enhancing potential impact of tsunami. Hence, based on the geomorphic features of the study area (Table 2) and their characteristics we have assigned vulnerability level (*after* Imamura, 2004) for each geomorphic feature (Table 3) with which tsunami vulnerability assessment for the study area was made (Table 4).

Table 2. Description of Beaches based on their geomorphic features

Description	Beach
Upland area	Chinna Muttom, Kanyakumari, Muttom.
Open coast	Ovari, Idinthakarai, Kuttapuli, Vattakottai, Vaniakudy, Lakshimipuram, Pallam, Mandaikadu, Midalam, Enayam.
Estuary zone	Perumanal, Manakudy, Rajakkamangalam, Kadiapattinam, Colachel, Taingapatnam.

Table 3. Criteria of Tsunami Hazard Classification

Zone	Description	Tsunami Hazard Category (Based on Inundation Extent)		
		High (m)	Medium (m)	Low (m)
Open Coast	Everywhere within 2km of coast	301 – 400	201 – 300	0 – 200
Coastal Estuary	Low lying flat topography of river valleys and bottomlands	501 – 750	251 – 500	0 - 250
Upland	All other areas inland of open coast zone	201 – 300	101 – 200	0 – 100

Table 4. Tsunami Hazard Classification along the Study Area

S.	Hazard Category	Beaches of corresponding hazard category
1.	Low	Ovari, Perumanal, Navaladi, Rajakkamangalam, Mandakadu, Kadiapatanam, Vaniakudy, Enayam, Taingapatnam
2.	Medium	Idinthakarai, Kuttapuli, Vattakottai, Lakshmipuram, Chinna Muttom, Midalam
3.	High	Kanyakumari, Manakudy, Muttom, Pallam, Colachel

Classification of the study area into different zones based on their geomorphic features and the relative inundation extent was attempted to delineate the boundaries of different categories of hazard levels. Tsunami hazard area boundaries are initially defined for each zone as enlisted herein based on elevation and inundation distance.

Open Coast Zone : This zone is directly exposed to the ocean. It includes all areas within 2 km distance across the coast. This area is vulnerable to inundation and high velocity waves.

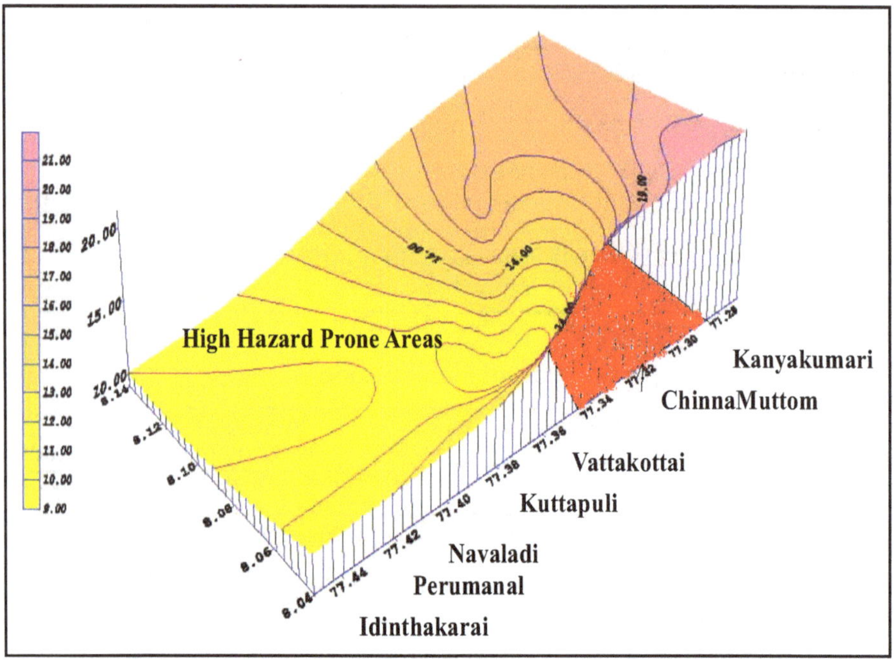

Figure 3a. Digital Elevation Model (DEM) from Ovari to Kanyakumari

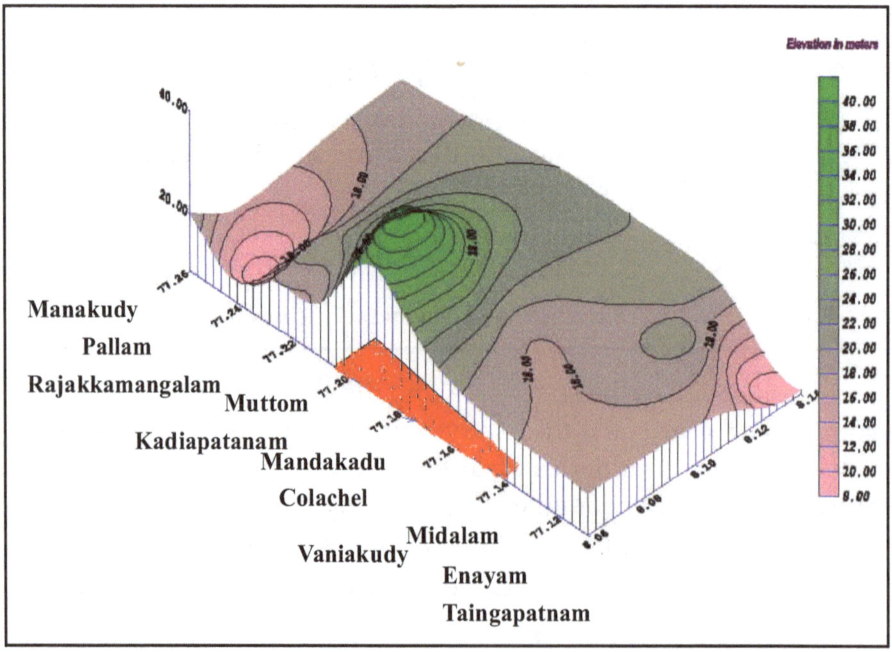

Figure 3b. Digital Elevation Model (DEM) from Manakudy to Taingapatnam

Coastal Estuary Zone : This zone is from the end of the Open Coast Zone to inland elevations of upto 20 - 30 m. This area is vulnerable to incursion of tsunami through river channels. Flooding potential is strongly dependent on ambient tide and water levels.

Upland Zone : All areas (except bays or estuaries) more than 2 km inland from the coast. This zone is not vulnerable to tsunami hazards.

It is to be noted that, this classification is appropriate for the field characteristics of the study area and may not be applicable elsewhere.

An integrated Tsunami Hazard Classification map has been prepared (Fig. 4) and it is inferred from this figure and the field and secondary data that the impact of tsunami surge was high in the southernmost part of the study area as most of the highly vulnerable coastal zone falls on that region namely, Kanyakumari, Manakudy, Muttom, Pallam and Colachel. Prime reason for this nature is that this zone has lowest elevation with sandy beaches, shows the absence of natural barriers and presence of scanty vegetation cover (either natural or plantation).

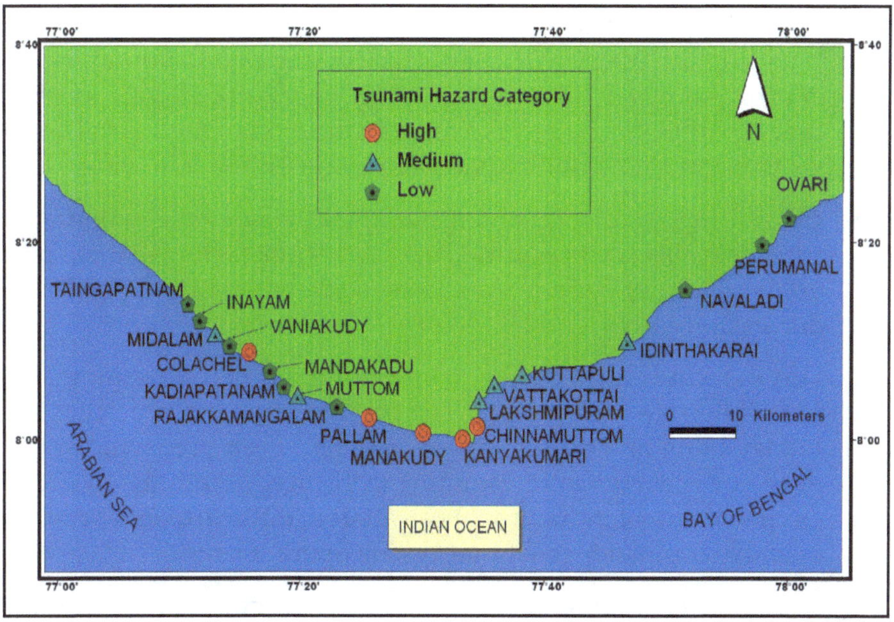

Figure 4. Integrated Tsunami Hazard Classification Map of the Study Area

Colachel was the only beach that suffered maximum destruction in the northwestern coast as the inundation was encouraged by the presence of river mouth. There are many other beaches neighboring river mouth but were not much affected due to the fact that the topography and geomorphic features of those regions are not conducive to inundation. Manakudy, due to its location in the southern tip of the continent facing the direction of the refracted waves from Sri Lanka facilitated inundation through estuarine channel and had suffered severe destruction. The presence of a promontory at Muttom had acted as a safeguard against tsunami waves.

The coastal zones of Idinthakarai, Kuttapuli, Vattakottai, Lakshmipuram, Chinna Muttom and Midalam are classified into medium hazard zones, as they were not much affected by the tsunami waves. The coastal zones of Ovari, Perumanal, Navaladi, Rakkamangalam, Kadiapatanam, Mandakadu, Vaniakudy, Enayam and Tangapatnam are classified into least vulnerable zones principally due to their highest topography compared to mean sealevel, geomorphic features and coastal vegetation.

CONCLUSIONS

Systematic survey and analyses of geological, geomorphological, oceanographic and remotely sensed multi-temporal and multi-source data in addition to interviews conducted among population living in a region that shows diverse topographic, vegetational, geomorphic and oceanographic characteristics for classification of the coastal zone towards tsunami vulnerability had demonstrated the following.

- Coastal topography, geomorphology, lithology, vegetation and presence/absence of natural barriers are the factors that control vulnerability of coastal zone to tsunami hazard in the order of decreasing importance.
- Having understood the prime controls of tsunami wave inundation, the study area had been categorized according to vulnerability and an integrated map had been prepared. In the event of recurrence of tsunami in future, assuming that the studied parameters have not been altered, this map could provide valuable inputs for relief and rescue operations.
- It is suggested that similar studies, covering other areas of Indian coast incorporating many other critical parameters such as soil, topographic information at decimeter scale, etc, have to be made in addition to simulation of tsunami surges at varying intensities for effective coastal zone management.

References

Angusamy, N. and Rajamanickam, G.V., 2000 Distribution of heavy minerals along the beach from Mandapam to Kanyakumari, Tamilnadu. Jour.Geol.Soc.India. v.56. pp.199-211.

Bandibas, J., Warita, K. and Kato, H., 2003 Interactive presentation of geological hazard maps using Geohazardview. Jour.Nat.Disast.Sci. v.25. pp.75-83.

Besana, G.M., Ando, M. and Mirabueno, H.M., 2004 The May 17, 1992 Event: Tsunami and coastal effects in eastern Mindanao, Philippines. Sci.Tsun.Haz. v.22. pp. 61-68.

Chandrasekar, N., 2005 Tsunami of 26th December 2004: Observation on inundation, sedimentation and geomorphology of Kanyakumari coast, south India. Proc. 22nd Internat.Tsunami Symposium, Greece. pp.22-28.

Chandrasekar, N. and Loveson Immanuel, J., 2005. GIS Supported categorisation of tsunami experienced beaches along the southeastern coast of India: Usage in mitigation activities. Proc.Nat.Sem.GIS Application in Rural Development, Hyderabad. pp.349-362.

Emery, K.O., 1961 A Simple method of measuring beach profiles. Limnol.Oceanogr. v.6. pp. 60.

Imamura, F., 2004 Risk Evaluation and Real Time Information for Tsunami Disaster Mitigation. Proc. Wave, Tide Observations and Modelings in the Asian-Pacific Region. pp.93-100.

Jaffe, B.E. and Gelfenbaum, G., 2002 Using Tsunami deposits to improve assessment of tsunami risk. Proc.Solutions to Coastal Disasters'02, ASCE. pp.836-847.

Keating, B., Whelan, F. and Brock, J.B., 2004 Tsunami Deposits at Queen's Beach, Oahu, Hawaii–Initial Results and Wave Modelling. Science of Tsunami Hazards. v.22. pp.23–43.

La Fond, E.C. and Prasada Rao, R., 1954 Beach erosion cycles near Waltair on the Bay of Bengal. Mem.Oceanogr. v.1. pp.63-77.

Loveson, V.J., Angusamy, N. and Rajamanickam, G.V., 1996 Usefulness of identifying different geomorphic blocks along the coast of southern Tamilnadu. Indian.Jour.Geomor. v.1. pp.97-110.

Minoura, K., Nakaya, S. and Uchida, M., 1994 Tsunami deposits in a lacustrine sequence of the Sanriku coast, northeast Japan. Sediment.Geol. v.89. pp.25-31.

Raval, U., 2005 Some factors responsible for the devastations in Nagapattinam region due to Tsunami of 26th December 2004. Jour.Geol.Soc.India. v.65. pp.647–649.

Shuto, N., 2001 Tsunami induced topographical change recorded in documents in Japan. Proc. ITS 2001, Session.3. No. 3-6. pp.513-522.

Walsh, T.J., Caruthers, C.G., Heinitz, A.C., Myers, E.P, Baptista, A.M., Erdakos, G.B. and Kamphaus, R.A., 2000 Tsunami hazard map of the southern Washington coast : Modelled Tsunami Inundation from a Cascadia subduction zone earthquake. Geologic Map GM-49. Report of Washington Division of Geology and Earth Resources. 12 p.

❑❑❑

6

Earthquake - The Creator of Geohazards

G. SATHISH
Department of Geology, Periyar University, Salem - 636 011.
E-Mail: <sathish.geol@gmail.com>

This chapter is an attempt to introduce basic facts about earthquake generation and the role of earthquakes in triggering secondary geohazards. Earthquake, unlike many other geohazards is yet to be fully understood towards occurrence and magnitude for better mitigation measures. Though unpredictibility of this hazard makes it dangerous, earthquakes seldom kill lives. Instead, poor ability of man-made structures against ground-shaking waves has been the real cause of life loss and damage to property in most of the earthquake incidences. Furthermore, earthquakes generate secondary geohazards such as landslide, failure of dams and reservoirs, etc. Thus, earthquake, as a natural hazard has its own characteristic way of potential impact and hence classifying vulnerability of different regions and creating awareness on potential impact among common public is essential.

INTRODUCTION

The term "earthquake" denotes "shaking/vibration/trembling/quivering/shuddering/ wobbling of Earth". Earthquake is the quickest and most unpredictable natural hazard. It is nothing but a series of suddenly generated elastic waves occurring in the Earth due to internal and/or external causes from shallow (few kilometers) to about 700 kilometers depth. It is a form of motion of wave energy transmitted through the surface layer (crust) of the Earth. One generalization that can be made out of scientific and mainstream media information about earthquakes is that, "earthquakes seldom kill; but the associated phenomena such as landslides, flooding, volcanic eruptions and collapse of natural and man-made structures cause the real damage to lives and property". This chapter is an attempt to describe the basic elements of earthquakes and few mitigation measures.

GENESIS AND PROPERTIES OF EARTHQUAKES

As discussed in few other chapters in this book, the Earth is made up of many plates floating on molten to semi-molten rock. At few places these plates collide and in few other places they move apart. When the Earth's plates move against each other, stress is put on the lithosphere. When this stress is great enough to reach critical level (unique to the lithology, structural characteristics, P-T conditions, etc. of the place) the lithosphere breaks or shifts. When it breaks, the stress is released as energy, which moves through the Earth in the form of seismic waves, which are called earthquakes. Sudden slip movements along faults produce most earthquakes. During movement, a large quantity of energy is instantaneously released, which shakes the ground. In the case of a transform fault, on which movement of plates is in a horizontal direction, slow bending of the rock that precedes the shock takes place over many thousands of years. Sometimes a slow, steady displacement known as fault creep occurs, which tends to reduce the accumulation of stored energy. Earthquakes can also be produced by volcanic activity, when magma rises or recedes within a volcanic chamber.

The point at which the stored energy in the form of seismic wave is released is called the "focus". "Epicenter" is the point on the surface of the Earth's crust directly above the focus. From focus, the energy is transmitted in the form of circular waves/seismic waves. Two types of seismic waves namely, *body waves* and *surface waves* are produced during earthquakes. Body waves travel through the interior of the Earth. They follow ray paths bent by the varying density and modulus (stiffness) of the Earth's interior. The density and modulus, in turn, vary according to temperature, composition and phase. Body waves transmit tremors of an earthquake. There are two kinds of body waves: primary (P) and secondary (S).

"P" waves are longitudinal or compression waves, which mean that the ground is alternately compressed and dilated in the direction of propagation. In solids, these waves generally travel almost twice as fast as S waves and can travel through any type of material. In air, these waves take the form of sound waves and hence they travel at the speed of sound. Typical speeds are 330 m/s in air, 1450 m/s in water and about 5000 m/s in granite. They are less destructive than the S waves and surface waves that follow them, due to their smaller amplitudes. "S" waves are transverse or shear waves, which mean that the ground is displaced perpendicularly to the direction of propagation. In the case of horizontally polarized S waves, the ground moves alternately to one side and then the other. S waves can travel only through solids, as fluids (liquids and gases) do not support shear stresses. Their speed is about 60% of that of P waves. S waves are several times larger in amplitude than P waves.

Surface waves are analogous to water waves and travel just under the Earth's surface. They travel slower than body waves. Because of their low frequency, long duration and large amplitude, they are the most destructive type of seismic waves. There are two types of surface waves: Rayleigh waves (R) and Love waves (L). "R" waves are surface waves that travel as ripples similar to those on the surface of water. They are slower than body waves, roughly 70% of the velocity of S waves. "L" waves are surface waves that cause horizontal shearing of the ground. They travel slightly faster than "R" waves, about 90% of the S wave velocity.

Earthquake with highest intensity in a location at a particular time is termed "Main Shock", the minor shocks prior to the main shock are termed as "Foreshocks" and the shocks that follow the main shock are termed as "After shocks". Based on depth of focus, cause and intensity, earthquakes are classified. Depending on depth of focus, earthquakes are classified into "shallow" (focus located within 60km depth from surface), "intermediate" (focus within 60 to 300km below surface) and "deep seated" (focus within 300 to 700km below surface).

Earthquakes are classified into class A, B, C, D and E. While class E is inclusive of all earthquakes of magnitude less than 5.3 and does not pose severe threat to engineering structures, class A is inclusive of all seismic events with >7.8 magnitude which can cause severe damage. The earthquakes that emanate from plate movements and faults are tectonic earthquakes. All others caused by volcanism, landslide and atomic explosion are non-tectonic.

MEASUREMENT OF EARTHQUAKES

Earthquakes are measured through many scales. The first is to describe the earthquake's intensity. Intensity is the measure of damage to the surface and the effects on humans. It records only observations of effects on the crust, not the actual ground motion or wave amplitude, which can be recorded by instruments. Seismometer is an instrument that measures and records motions of the ground generated by earthquake and other man made nuclear explosions. Seismogram is the record of ground vibrations. The records of wave amplitude and distance of travel define magnitude of an earthquake. Commonly used scale of magnitude is the Richter scale. To determine the Richter magnitude, information collected by seismometers is used. Using a seismogram, the time difference between the recording of the P wave and the S wave is determined and matched to a corresponding distance value. The single maximum amplitude recorded on the seismogram is calculated and a line is drawn between the amplitude scale and the distance scale. The line crosses another scale, which corresponds to the magnitude. This magnitude scale is logarithmic; meaning each step in magnitude is exponentially greater than the last. The Richter magnitude scale is presented herein (Table 1). In addition to Richter magnitude, body wave magnitude and surface wave magnitude are also measured. These magnitude scales differ by the type of wave amplitude that is measured from the seismogram and the mathematical formula used to determine the magnitude. They are all, however, logarithmic scales.

Mercalli scale (Table 2) is another method of measure of earthquake's intensity based on subjective observations on earthquake in terms of lives lost and damage to structures. It is useful in assessing the impact based on post event field observations.

Table 1. Richter scale of magnitude of earthquake

Richter scale	Description	Characteristics
Less than 2.0	Micro	Micro-earthquakes, not felt.
2.0-2.9	Minor	Generally not felt, but recorded.
3.0-3.9	Minor	Often felt, but rarely causes damage.
4.0-4.9	Light	Noticeable shaking of indoor items, rattling noises. Significant damage unlikely.
5.0-5.9	Moderate	Can cause major damage to poorly constructed buildings over small regions.
6.0-6.9	Strong	Can be destructive in populated areas.
7.0-7.9	Major	Can cause serious damage over larger areas.
8.0-8.9	Great	Can cause serious damage in hundreds of sq.km. area.
9.0-9.9	Great	Devastating in areas several thousands of sq.km. area.
10.0+	Epic	Never recorded.

(Source: www.usgs.gov)

Table 2. Modified Mercalli Scale of earthquake intensity

Category	Descriptive scale
I. Instrumental	Not felt by many people unless in favorable conditions.
II. Feeble	Felt only by a few people at best, especially on the upper floors of buildings. Delicately suspended objects may swing.
III. Slight	Felt quite noticeably by people indoors, especially on the upper floors of buildings. Many do not recognize it as an earthquake. Standing motorcars may rock slightly. Vibration similar to the passing of a truck. Duration estimated.
IV. Moderate	Felt indoors by many people, outdoors by few people during the day. At night, some awakened. Dishes, windows, doors disturbed; walls make cracking sound. Sensation like heavy truck striking building. Standing motorcars rocked noticeably. Dishes and windows rattle alarmingly.
V. Rather Strong	A transient stage between moderate and strong types. Damage is sporadic and varies according to local conditions.

Contd...

Table 2. Contd...

VI. Strong	Felt by all; many frightened and run outdoors, walk unsteadily. Windows, dishes, glassware broken; books off shelves; some heavy furniture moved or overturned; a few instances of fallen plaster. Damage slight.
VII. Very Strong	Difficult to stand; furniture broken; damage negligible in building of good design and construction; slight to moderate in well-built ordinary structures; considerable damage in poorly built or badly designed structures; some chimneys broken. Noticed by people driving motorcars.
VIII. Destructive	Damage slight in specially designed structures; considerable in ordinary substantial buildings with partial collapse. Damage great in poorly built structures. Fall of chimneys, factory stacks, columns, monuments and walls. Heavy furniture moved.
IX. Ruinous	General panic; damage considerable in specially designed structures, well designed frame structures thrown out of plumb. Damage great in substantial buildings, with partial collapse. Buildings shifted off foundations.
X. Disastrous	Some well built wooden structures destroyed; most masonry and frame structures destroyed with foundation. Rails bent.
XI. Very Disastrous	Few, if any masonry structures remain standing. Bridges destroyed. Rails bent greatly.
XII. Catastrophic	Total damage - Almost everything is destroyed. Lines of sight and level distorted. Objects thrown into the air. The ground moves in waves or ripples. Large amounts of rock may move position.

(*Source:* USGS)

Another way to measure vertical and horizontal movement of the earth is through the use of GPS. GPS is navigation and positioning system developed by the Department of Defense, U.S.A. in the early 1970's. Among its many applications, scientists use GPS to monitor the movement of the Earth's crust, all over the World, between and during earthquakes. From these measurements, maps and models can be created to show how fast and in what direction the crust is moving due to both plate and fault movement. By using GPS and array of GPS networks, it will be possible in the near future to reduce damage, improve emergency preparedness and response by determining, in advance, areas at risk.

IMPACTS OF EARTHQUAKES

Earthquakes are invariably catastrophic in nature due to many reasons. First is that they "actually shake the very foundations over which all man made structures rest on. Second is that, these are the most unpredictable earth's processes in terms of their timing, place, intensity, etc. Third is that, their potential impact depends on too many factors such as velocity, period, local geology, nature and strength of man-made structures that could not be generalized for better understanding and mitigation measures. Fourth is that, earthquakes trigger few other geohazards namely, landslide, mass avalanche, volcanism and tsunami which in turn are catastrophic in their own rights. In addition to human life loss and collapse of infrastructure, earthquakes can render lands inhospitable and change landscape permanently. Table 3 lists few of the deadliest earthquakes in human history.

ZONES OF EARTHQUAKES

According to NASA website, almost 95% of earthquakes that occurred all over the World are confined to few seismic belts, aligned exactly along tectonic plate margins. These are the geographic regions wherein intense volcanism is also reported. The following list shows the active seismic belts of the World (*Source:* http://science.nasa.gov). Regions other than these are considered as non-seismic areas.

The Circum-Pacific Belt : 81% of major earthquakes occur in a continuous belt along the peripheral zone of Pacific Ocean and this region is called Circum-Pacific seismic belt. The Pacific Ocean is enclosed by different plate boundaries and volcanic activity is well pronounced along this belt. This belt extends from Aleutian Islands through Alaska, Coasts of North and South America, New Zealand, the East Indies, the Phillippines, Japanese islands and Kuril Islands. 99% of deep, 90% of intermediate and 80% of shallow earthquakes occur in this belt.

The Alpine belt : It is the Mediterranean trans-asiatic belt, running approximately in an east west direction, from the Azores to New Guinea where the belt joins with the Pacific belt. The Belt runs through Mediterranean, Himalayan region, Sumatra, Java and Timor. Intermediate and shallow earthquakes commonly occur in this belt.

Table 3. Deadliest earthquakes *(Source: USGS)*

Rank	Name	Date	Location	Fatalities	Magnitude	Remarks
1	"Shaanxi"	January 23, 1556	Shaanxi, China	830,000	8.0	
2	"Tangshan"	July 27, 1976	Tangshan, China	255,000 (official)	7.5	Estimated death toll as high as 655,000.
3	"Indian Ocean"	December 26, 2004	Off West coast of Northern Sumatra, Indonesia	~230,210 dead and missing	9.3	Deaths from earthquake and tsunami.
4	"Aleppo"	October 11, 1138	Aleppo, Syria	230,000	8.5	Death toll is disputed.
5	"Damghan"	December 22, 856	Damghan, Iran	200,000		
	"Gansu"	December 16, 1920	Ningxia-Gansu, China	200,000	8.6	Major fractures, landslides.
7	"Ardabil"	March 23, 893+	Ardabil, Iran	150,000		
8	"Great Kantô"	September 1, 1923	Kantô region, Japan	143,000	7.9	Great Tokyo fire.
9	"Ashgabat"	October 6, 1948	Ashgabat, Turkmenistan	110,000	7.3	
10	"Kashmir"	October 8, 2005	India & N.W.F.P, Pakistan	100,000 (estimated), 80,000 (official)	7.6 or 7.8	3.5 million people homeless, 100,000 feared dead

The central Asian Pamir-Baikal belt : It is a zone of shallow earthquakes. It extends northwardly from Pamir to Baikal of Russia.

Ridge belt : Earthquakes of shallow type occur frequently along the mid-ocean ridges. This belt includes Mid-Atlantic Ridge, Arctic and Indian Ocean Ridges.

The Rift Zone belt : It includes east African Rift Valleys. Shallow earthquakes characterize it. In addition to these prominent belts, the Central Pacific Ocean covering Hawaiin islands belt and the East of Alpine and Pamir-Baikal belts are other less significant seismic belts.

INDIAN SCENARIO

Indian sub-continent is geologically divided into few categories namely, the Himalayan highlands, Gangetic plains, Deccan Plateau and Peninsular India. While the Himalayan highlands and Gangetic plains are traditionally considered to be tectonically active and are prone to earthquakes, recent studies (Krishna Brahmam and Kochhar, 1989; Subramanian and Gopalakrishnan, 2002; O.P. Singh, *et al.* 2008, Vinod Kushwah *et al.* 2008; Prantik Mandal, and Chadha, 2008; Dipak Ghosh *et al.* 2007; Ulrich Kamp *et al.* 2008; Somenath Karmakar *et al.* 2008; Chandrani Singh et al. 2008; Jayangondaperumal and Thakur, 2008; Boschi *et al.* 2006; Purnachandra Rao *et al.* 2006; Das *et al.* 2006) have unequivocally indicated that no part of India is safe from earthquakes albeit differences in intensity (Fig.1) and depth of focus. According to these studies, the deltaic coastal regions located along eastern India are as vulnerable as that of Gangetic plains. In addition, neotectonic reactivation of older faults in Deccan Plateau and Peninsular India are also suggested to be seismically active. Thus, it could be stated that, as is elsewhere, Indian sub-continent is not safe from earthquakes. However, probable intensity and magnitude of earthquake vary regionally and most importantly influenced by local geological conditions. At this juncture, it would not be foolproof to classify the Indian sub-continent according to earthquake vulnerability on a regional scale. With the existing knowledge pertaining to earthquake, it is near impossible to predict it with permissible tolerance levels. Hence, formulation of nationally coordinated earthquake hazard zonation of Indian Territory at micro scale, incorporating local geological, population, landuse and other data should be made for developing workable and effective mitigation measures.

MITIGATION MEASURES

Having admitted that earthquake is one of the deadliest unpredictable geohazard and also that no part of the Indian subcontinent is safe from earthquake, it is imperative to establish mitigation measures. As is the case of any other geohazard, earthquakes could not be prevented and hence, learning to live with this natural geological phenomenon with minimal property and life loss is the objective of mitigation measures. This section presents few measures specific to Indian scenario.

- Foremost requirement is, mapping the country pertaining to seismic hazard at village level scale and creating awareness among local public.

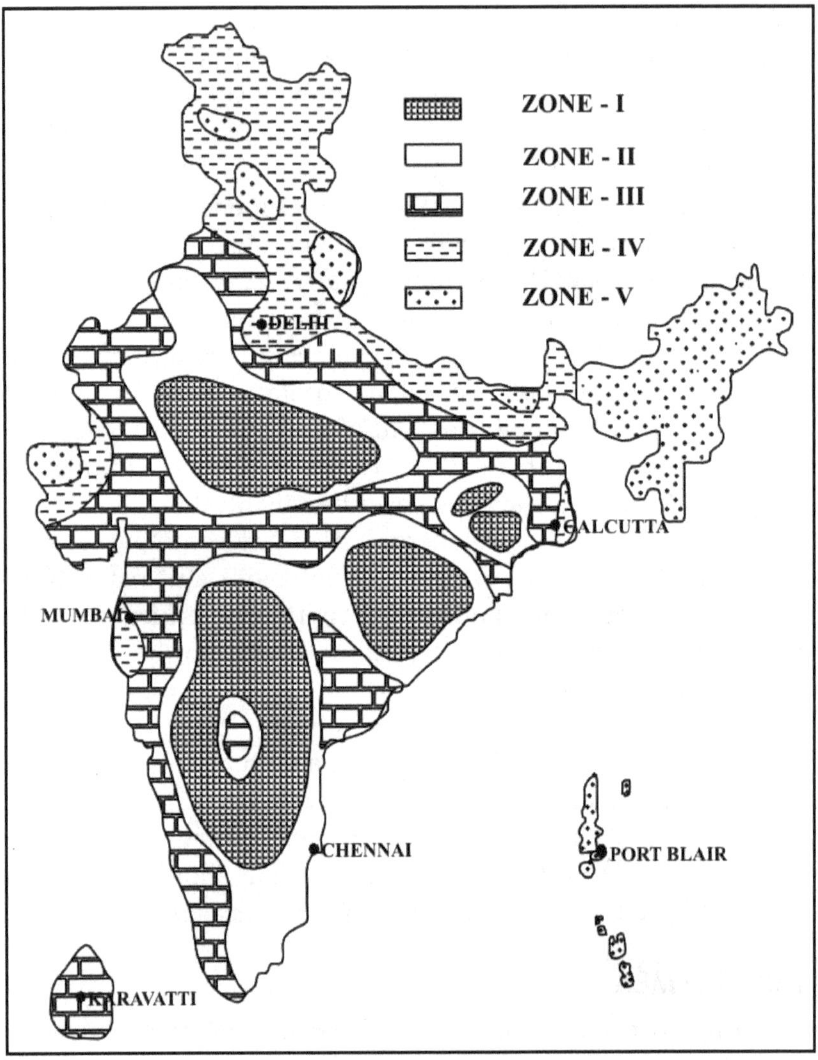

Figure 1. Earthquake hazard zonation map of India
(*Source:* National Disaster Management Authority)

Zone V : Covers the area liable to seismic intensity IX and above on Modified Mercalli Intensity Scale. This is the most severe seismic zone and is referred here as Very High Damage Risk Zone.

Zone IV : Gives the area liable to MM VIII. This zone is second in severity to zone V. This is referred here as High Damage Risk Zone.

Zone III : The associated intensity is MM VII. This is termed here as Moderate Damage Risk zone.

Zone II : The probable intensity is MMVI. This zone is referred to as Low Damage Risk Zone.

Zone I : Here the maximum intensity is estimated as MMV or less. This zone is termed here as Very Low Damage Risk Zone.

- As collapse of construction structures causing life loss during earthquakes has been the reason rather than earthquake itself, creation of awareness among civil engineers and masons about design and construction of earthquake resistant structures are important. In this regard, equipping them with requisite information is as much important as strict enforcement of guidelines.
- With the sophisticated network of seismic sensors linked with GPS and satellites available, creation of dense network of sensors spread all over the country is needed. This would not only help relay information on the occurrence of seismic event in an area, but also would help plan and implement relief and rescue missions. Although a beginning is made by organizations such as National Geophysical Research Institute, Hyderabad, creation of dense array of sensors, dynamic modeling and satellite networked relay systems for disseminating information to the people of vulnerable regions and appropriate decision centres are necessary.

References

Boschi, E., Casarotti, E., Devoti, R., Melini, D., Piersanti, A., Pietrantonio, G. and Riguzzi, F., 2006 Coseismic deformation induced by the Sumatra earthquake, Jour.Geodyn. v.42. pp.52-62.

Das, N.K., Choudhury, H., Bhandari, R.K., Ghose, D., Sen, P. and Sinha, B., 2006 Continuous monitoring of ^{222}Rn and its progeny at a remote station for seismic hazard surveillance. Radiat.Measure. v.41. pp.634-637.

Ghosh, D., Deb, A., Sengupta, R., Patra, K.K. and Bera, S., 2007 Pronounced soil-radon anomaly-Precursor of recent earthquakes in India. Radiat.Measure. v.42. pp.466-471.

Iwan, W.D., 2006 (Ed.) Summary Report of the Great Sumatra Earthquakes and Indian Ocean Tsunamis of 26 December 2004 and 28 March 2005. EERI Publication-2006-06.

Jayangondaperumal, R. and Thakur, V.C., 2008 Co-seismic secondary surface fractures on southeastward extension of the rupture zone of the 2005 Kashmir earthquake. Tectonophys. v.446. pp.61-76.

Kamp, U., Growley, B.J., Khattak, G.A. and Owen, L.A., 2008 GIS-based landslide susceptibility mapping for the 2005 Kashmir earthquake region. Geomorph. v.101. pp.631-642.

Karmakar, S., Rathore, A.S., Kadri, S.M., Dutt, S., Khare, S. and Lal, S., 2008 Post-earthquake outbreak of rotavirus gastroenteritis in Kashmir (India): An epidemiological analysis. Pub.Heal. v.122. pp.981-989.

Krishna Brahmam, N. and Kochhar, N., 1989 Tosham area, India: Petrographical, chemical, and gravity studies. Proc.Inter.Symp.Structure and dynamics of the Indian Lithosphere, National Geophysical Research Institute, Hyderabad. pp.110-115.

Kushwah, V., Singh, V. and Singh, B., 2008 Ultra low frequency (ULF) amplitude anomalies observed at Agra (India) and their association with regional earthquakes. Phys.Chem.Earth. pp.(in press).

Mandal, P. and Chadha, R.K., 2008 Three-dimensional velocity imaging of the Kachchh seismic zone, Gujarat, India. Tectonophys. v.452. pp.1-16.

Radhakrishnan. V., 1996 General Geology, Second edition, V.V.P. Publishers, Tuticorin. 280p.

Rao, P.N., Kumar, P., Kalpna, K., Tukuda, T. and Ramesh, D.S., 2006 The devastating Muzaffarabad earthquake of 8 October 2005: New insights into Himalayan seismicity and tectonics. Gond.Res. v.9. pp.365-378.

Rao, P.N., Kumar, P., Kalpna, K., Tukuda, T. and Ramesh, D.S., 2006 The devastating Muzaffarabad earthquake of 8 October 2005: New insights into Himalayan seismicity and tectonics. Gond.Res. v.9. pp.365-378.

Singh, P., 2004 Engineering and General Geology. VII revised edition. S.K. Kataria & Sons, New Delhi. 580p.

Singh, C., Ramana, D.V., Chadha, R.K. and Shekar, M., 2008 Coseismic responses and the mechanism behind M_W 5.1 earthquake of March 14, 2005 in the Koyna–Warna region, India. Jour.Asian Earth Sci. v.31. pp.499-503.

Singh, O.P., Chauhan, V., Singh, V. and Singh, B., 2008 Anomalous variation in total electron content (TEC) associated with earthquakes in India during September 2006–November 2007. Phys.Chem.Earth. pp.(in press).

Stein, S. and Mazzotti, S., 2007 Continental Intraplate Earthquakes: Science and Policy Issues. Geol.Soc.Amer.Spl.Paper 425.

Strahler, A. and Strahler, A., 1997 Physical Geography, John Wiley and Sons, U.S.A. 567p.

Subramanian, K.S. and Gopalakrishnan, K., 2004 Earth tremors in the Coastal Belt of Tamil Nadu and Pondicherry in September, 2004 - A geological analysis. Jour.Geol.Soc.India. v.60. pp.691-694.

USGS., 2008 www.usgs.gov. List of World's Largest Earthquakes.

USGS., 2008 www.usgs.gov. The Richter Magnitude Scale.

USGS., 2008 www.usgs.gov. "M7.5 Northern Peru Earthquake of 26 September 2005" (pdf).

www. Wikipedia.com

❑❑❑

7

Earth Tremor in Jind Region, Central Haryana : Seismogenesis and Mitigation

G. VALLINAYAGAM, L.GOPESHWOR SINGH & NARESH KUMAR
Department of Geology, Kurukshetra University, Kurukshetra - 136 119.
E-mail: <gvallinayagam@rediffmail.com>

This chapter presents a case of Jind region of Haryana State, Northern India which had experienced mild tremors of ~ 3 on Richter scale magnitudes in the recent past with a gap of 1 to 2 years. Rejuvenation of the lineament controlled volcanoseismic process is identified as the probable cause of these tremors. This chapter describes seismotectonic characters, micro and mega seismicity of the area and few mitigation measures.

INTRODUCTION

The Haryana state consists of Delhi Super Group (DSG), Indo-Gangetic alluvial plain in the form of alluvial fan, sand dune, alluvium of Ambala, Yamuna and Ghagghar rivers and Siwalik Super Group (SSG) in the southern, central and northern parts respectively (Fig.1). The study area Jind is located in the central part of Haryana and comprises 200 - 600 m thick Indo-Gangetic alluvium. The alluvium is of Quaternary Period and composes alternating layers of clay, silt, sand and kankar. The Bhiwani and Tosham areas, located south of

Jind, show the occurrence of Neoproterozoic granite and rhyolite of Malani Igneous Suite (MIS). The rocks occur in the form of ring structure/ring dyke on the hillocks amidst the alluvium. The metasediments (mica schist and quartzite) of Delhi Super Group are underling MIS at Tosham (located at 70 km southwest of Jind). As Jind region experiences mild tremors and there is no previous published work pertaining to the geology, tectonics and seismicity of the area the chapter records these aspects for better understanding and mitigation of earthquake hazard.

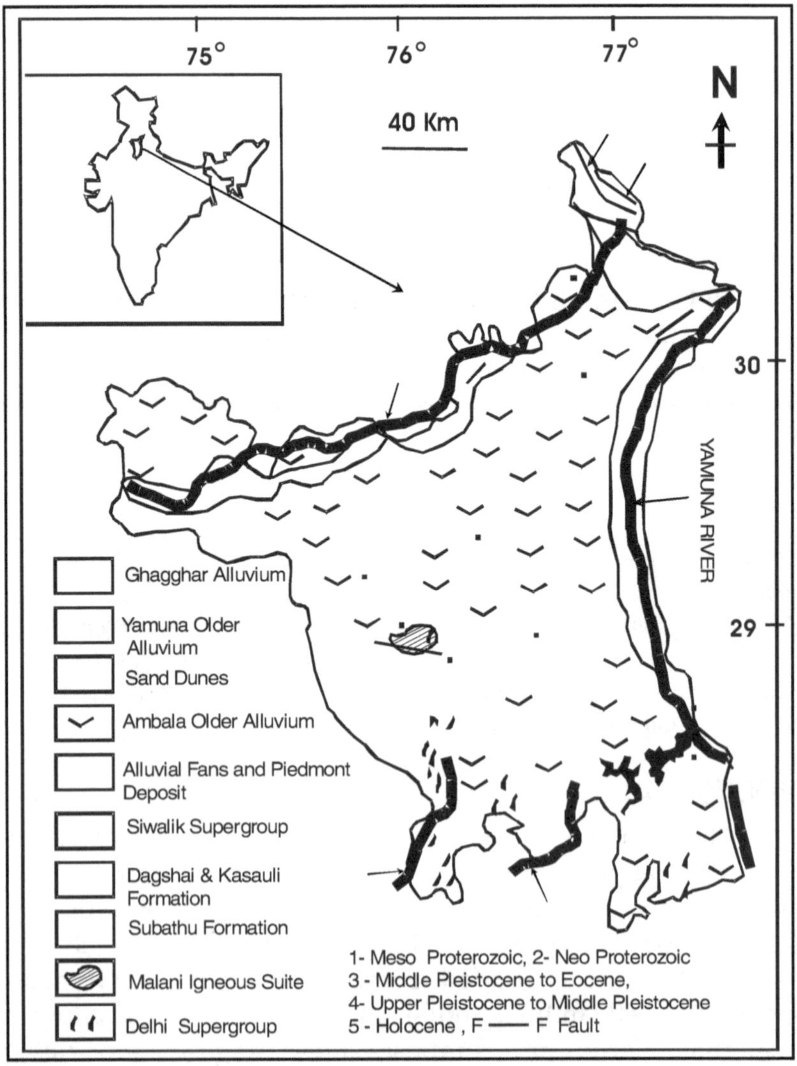

Figure 1. Geology and Geomorphology of Haryana (*after* Thussu, 1999)

SEISMOTECTONICS

The field investigations were carried out to assess the geological setting in and around Jind area (~300 km^2), wherein tremors were felt. The area is covered with alluvium, which is composed of unconsolidated sediments such as clay, sand and silt with variable thickness that extends more than 100 meters as evidenced through borehole cores. Occurrence of basement rock at about a depth of ~200m is observed (Fig. 2).

Older Alluvium with numerous earth mounds is exposed in the area. The height of earth mound near Kinana (most affected area ~ 7 km SE of Jind) is 18 meters high with gentle slope and orientated towards northeast-southwest direction. The number of mounds in the southern areas of Jind is more as compared to northern areas. Surface rock exposures are not present in the region. However, volcanoplutonic rocks and metasediments belonging to MIS and DSG respectively, are exposed in Tosham area. Magmatic episodes in Tosham are characterized as subvolcanic, alkaline, anorogenic, High Heat Producing (HHP) type formed due to the lineament controlled hotspot tectonics. Rocks of Tosham occur on triple gravity junction and a considerable portion of gravity low is caused by acute granite intrusion (Fig. 3). Hence, it is inferred that strike of Aravalli range turns northwest from Tosham. Tosham gravity junction is flanked by the Aravalli-Delhi basement towards Himalaya, in the form of Delhi-Haridwar ridge and Delhi-Moradabad ridge in the eastward direction and Delhi-Lahore ridge in the NW direction. Tosham magmatic rocks have recorded highest heat flow of 96 ± 3.2 mWm^{-2} within the Delhi mobile belt (Sunder *et al.* 1990) and attributed to the high concentration of the radioactive elements in the rocks exposed in the area. Kochhar *et al.* (1991) reported U 97 ppm and Th 111 ppm in the Tosham granites and 0.6 wt% of U in the zircon mineral separated from the Tosham granite. The heat flow from the upper mantle is generally responsible for the high surface heat flow in the region. The high heat acts a 'steam engine', which prolongs the fluid circulation in the crust. Incidentally, both Tosham and Jind areas lie along the same lineament, which is called as Jhunjhunu lineament. Jhunjhunu is located ~150 km southwest of Tosham and the exposed rocks are volcanoplutonics and included in the MIS (Kochhar *et al.* 1991).

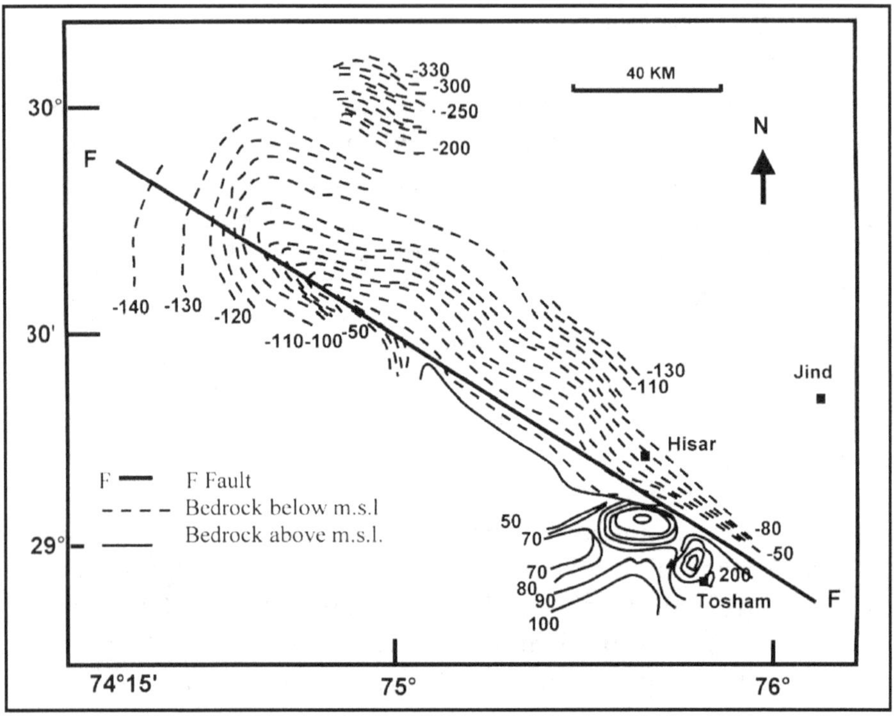

Figure 2. Isopach map of depth to basement rock in and around Tosham- Jind areas based on borehole data (*modified after* Thussu, 2006)

MICRO AND MACRO SEISMICITY

The data pertaining to the earth tremors of Jind region have been drawn through field investigations, interactions with local populations, search from various media and websites like Amateur Seismic Centre (www.asc-india.org) and India Meteorological Department (IMD). Earth tremors of 1.9 to 3.5 magnitudes in series were felt in the Jind city and its surrounding area during November 2003 – January 2004. There were about 15 tremors with a gap of few days. In addition, each major tremor was followed by less intense after shocks within an hour or so. The largest event (3.5 Magnitude) in the area was on 23rd December 2003 at 1.54 hours and experienced by entire Jind District as well as the adjoining Hisar and Kaithal originated at an epicenter 12 km SE of Jind (Latitude 29° 235′ N; Longitude 76° 402′ E) at a depth of about 4.6 km. Latest event occurred on 2nd March 2008 at 13.13 hours and lasted for 20 seconds. There are many other tremors that were felt by the people of this region and reported by local media on 19th March, 27th July 2004, 8th February and 27th July 2005. The highest magnitude (3.5) occurred in the area has probably caused the development of minor fissures on the earth, cracks on the buildings and tube wells and damaged wall plasters. Jind is one of the fertile agricultural areas of the state and its tube wells are the main source of water for irrigation, drinking and domestic purposes. There are thousands of tube wells with an average depth of 500 feet with negligible recharging. Some of them are yielding hot water with high temperature (up to 45 degree Celsius) and they are located around 40 km NE of Jind region. These observations suggest heating of groundwater probably by underground volcanic activity. In this context, occurrence of mild earth tremors in the regions should be taken seriously, before the occurrence of any catastrophic event.

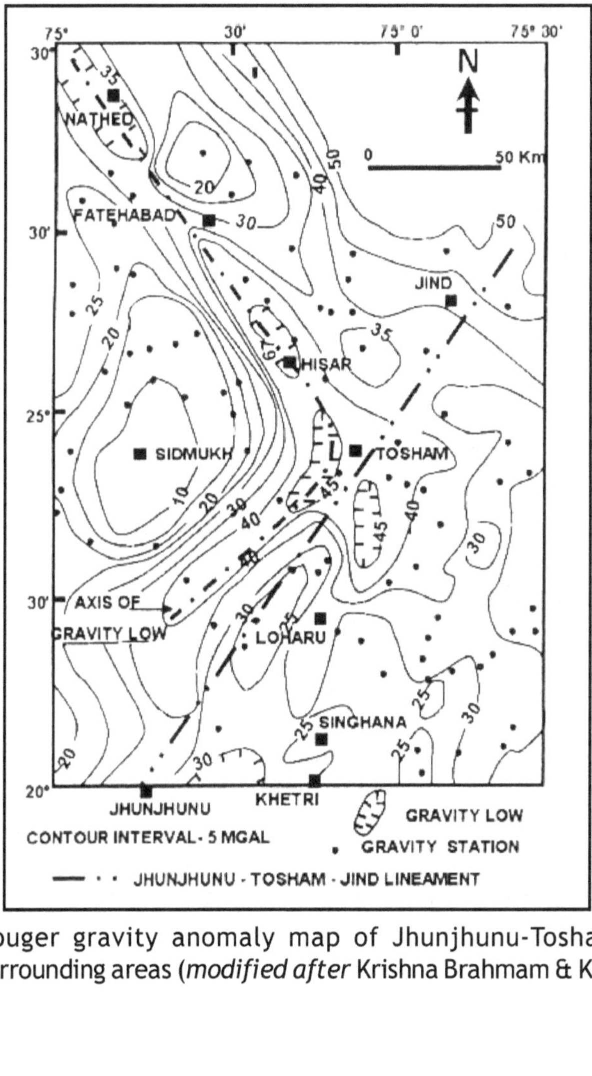

Figure 3. Bouger gravity anomaly map of Jhunjhunu-Tosham-Jind and surrounding areas (*modified after* Krishna Brahmam & Kochhar, 1989)

SEISMOGENESIS AND MITIGATION MEASURES

The study area is located in the seismic zone IV in seismic atlas of India. This information, coupled with the fact that there are many active hot water bearing tube wells in the area, suggests volcano induced seismic events in the area. As 200m thick alluvial deposits cover the area, any major earth tremor would devastate the entire region similar to the Latur earthquake. Present studies in the areas suggest that there is a tectonomagmatic relationship along the Jhunjhunu lineament on Jhunjhunu-Tosham-Jind corridor. Authors of this chapter feel that reactivation of lineament due to active crustal forces related to the MIS magmatic episodes is probable cause of these mild seismic activities in Jind region. The felsic rocks of the Jhunjhunu-Tosham are formed by the low degree partial melting of sialic rock in the earth crust. There have been atleast two abortive attempts by Indian lithosphere to rift, one related to Aravalli-Delhi mobile belt at 1500-1100 Million Years ago (Gopalan and Choudhary, 1984) and other related to the major tectonomagamtic event of MIS at 750 Million Years ago. The seismotectonic data supports the relationship between volcanic and seismic activities. Instrumentally recorded largest earthquake (6.0 magnitude; 15.58 hours; 28.2N:77.4E) in Haryana state was during August 1960 at Gurgaon-Faridabad. In Jind region, the earliest and largest event is 23rd December 2003. No other historical earth tremor reports are traceable in the region. The Indo-Gangetic basin is bounded by many active faults and covered with huge thickness of alluvium. Hence they are under tremendous stress and instability for millions of years. Due to the reactivation of the tectonomagmatism that is energized by crustal processes, the study area could be termed as "high risk" area prone to volcanic and seismic hazards. These inferences are supported by the views of Subramanian and Gopalakrishnan (2002).

Damage caused by earth tremors and associated impacts on life and infrastructure of an area is depending on various geological, geomorphological and lithological factors in addition to strength of building and other structures, timing of the event and awareness about potential impact among general public and readiness of administrative machinery. Continuous monitoring of the area in respect to physical, chemical and biological changes within the rocks/sediments and water is necessary. In addition, educating the general public pertaining to earthquake resistant building, relief and rescue measures is to be taken to minimize the potential loss.

References

Gopalan, K. and Choudhary, A.K., 1984 The crustal records in Rajasthan. Proc.Indian Acad. Sci. (Earth and Planetary Sci.). v.93. pp.337-342.

Kochhar, N., Vallinayagam, G. and Gupta, L.N., 1991 Zircons from the granitic rocks of the Malani Igneous Suite: Morphological and chemical studies. Jour.Geol.Soc.India. v.38, pp.561-576.

Kochhar, N., Sharma, R. and Vallinayagam, G., 1991 A note on Tin bearing potential of the granites of Jhunjhunu area, District Jhunjhunu, Rajasthan. Ind.Miner. v.45. pp.99-104.

Krishna Brahmam, N. and Kochhar, N., 1989 Tosham area, India: Petrographical, chemical, and gravity studies. Proc.Inter.Symp.Structure and dynamics of the Indian Lithosphere, National Geophysical Research Institute, Hyderabad. pp.110-115.

Subramanian, K.S. and Gopalakrishnan, K., 2004 Earth tremors in the Coastal Belt of Tamil Nadu and Pondicherry in September, 2004 - A geological analysis. Jour.Geol.Soc.India. v.60. pp.691-694.

Sunder, A., Gupta, M.L. and Sharma, S.R., 1990 Heat flow in the Trans-Aravalli Igneous Suite, Tosham, India. Jour.Geodyn. v.12. pp.89-99.

Thussu, J.L., 1999 Geoenvironmental appraisal of Haryana. Jour.Geol.Soc.India. v.54. pp.621-640.

Thussu. J.L., 2006 Geology of Haryana and Delhi. Text Book Series Geol.Soc.India. 61p.

❏❏❏

8

Volcanoes, Volcanism and Mitigation

A.KARTHIKEYAN
Department of Geology, Periyar University, Salem – 636 011.
E-mail: <akarthi_2006@rediffmail.com>

This chapter introduces the readers about volcanoes and associated features and hazards. Volcanic disasters are like any other geohazards in causing dangerous impacts. However, volcanic explosions are preceded by many surficial expressions that can be detected through satellite imageries, thermal sensors, change in magnetic and gravity anomalies, disappearances of vegetation, sudden appearances of anomalous water vapor, thermal springs and abnormal behavior of animals. Hence, unlike other geohazards, volcanic hazards could be forecast with fairly accurate time limits that may allow evacuation and precautionary safety measures. While avoidance is the best mitigation measure, proper land use planning, provision of scientifically designed damming and diversion methods could save lives and infrastructure from lava flows. Though, atmospheric contamination by volcanic gases, ash cloud and pyroclastic materials could not be thwarted and controlled and thus they could cause havoc.

INTRODUCTION

The term "Volcano" is derived from the name of Roman mythical God of fire "*Vulcan*". A volcano is an opening of different shapes, through which, molten rock and gas get ejected either violently or simply flow over earth's surface. The molten rock and gaseous materials are drawn from deep inside of crust or mantle of the Earth. Repeated ejection of molten materials through vent, fissures or plate margins forms volcanic mountains on earth's continental or submarine regions.

Volcanism is a threat not only to the lives and infrastructure, but it also drastically alters landscape, impacts atmospheric quality and renders forests, cities and villages and cultivation lands useless. In addition, volcanic eruptions are capable of causing changes in global atmospheric cycles. For example, the eruptions of El Chicho'n in Mexico in 1982, Laki volcano in Iceland and Asama Volcano in Japan in the year 1783 and Krakatau volcano of Japan in the year 1883 have all impacted global climate and atmorpheric dynamics.

A volcanic risk is defined in terms of the potential hazard produced by an eruption, degree of vulnerability of people and infrastructures and value of the territory exposed by future eruptions. Although volcanic hazard could not be predicted, as is the case of many other geohazards, observations in and around known volcanic regions could provide a fairly accurate prediction. Hence, evacuation and rescue efforts could be made atleast to some extent as far as the volcanic hazard is concerned. Despite this, by their very nature of triggering many other geohazards such as earthquakes, landslides, atmospheric pollution and mass failure, volcanoes are no less evil and thus are dangerous. This chapter details basic aspects of volcanism, volcanoes, volcanic materials and mitigation measures of volcanic hazard.

ORIGIN OF VOLCANOES

The Earth is made up of solid plates called "crust" that float over molten magma of asthanosphere and mantle (Sathish, *this volume*). The molten magma is under pressure and finds weak spots in the crust and gets molten material, gas and rock fragments ejected through them. The plates that float over molten magma have two different types of motions. One is that plates that move against each other (Converging Plate Margins - CPM) and the other type is that plates that move away from each other (Diverging Plate Margins - DPM). While the collisional movement along CPM melts the subducted earth portion to produce lava which in turn gets ejected via least pressure zones, forming volcanic chain, the plate margins from where plates move away from each other exposes weak spots through which new surface forms also becomes active volcanic region. Thus, volcanic mountain chains occur in three different regions, distributed in continental and submarine regimes, origins of which differ considerably.

CLASSIFICATION OF VOLCANOES

Morphological Classification

Based on morphology, volcanoes are classified in to four broad categories namely, cinder cones (Fig.1a), composite volcanoes (Fig.1b), shield volcanoes (Fig.1c) and lava domes (Fig.1d). The cinder cones are the simplest of all volcanoes. They are made up of solidified lava flows ejected from a single vent. Cinder cones are usually circular or oval in shape. They have a bowel shaped crater. Most number of cinder cones is found to occur in North America. Composite volcanoes have many vents through which ejection of lava take place. However, the central vent is the major conduit from which many branching vents form and eject volcanic material. Composite volcanoes form alternating layers of lava flow, volcanic ash and other pyroclastic materials. Composite volcanoes are comparatively steeper than shield volcanoes and rise above 3000 meters. The craters of these volcanoes some times get filled with water or ice or solidified lava domes. Mount Fuji in Japan, Mount St. Helens in U.S.A., and Mt. Etna in Italy are of this type. The shield volcanoes are typified by highly viscous lava flows that usually "flow" slowly from a central vent or vents. This type of volcano never explodes violently. The volcanic materials get supplied from a basaltic chamber and the lava flows form gentle slopes of less than 10°. Mauna Loa of Hawaii is the largest shield volcano. Volcanic domes or lava domes are formed by relatively small, bulbous masses of lava. The lava is highly viscous and flows long distances. A dome grows largely by expansion as it grows its outer surface, cools and hardens, then shatters, spilling loose fragments down its sides. Some domes form craggy knobs or spines over the volcanic vent, whereas others form short, steep-sided lava flows known as "coulees." Volcanic domes commonly occur within the craters or on the flanks of large composite volcanoes. Mont Pelée in Martinique, Lesser Antilles, and Lassen Peak and Mono domes in California are examples of lava domes.

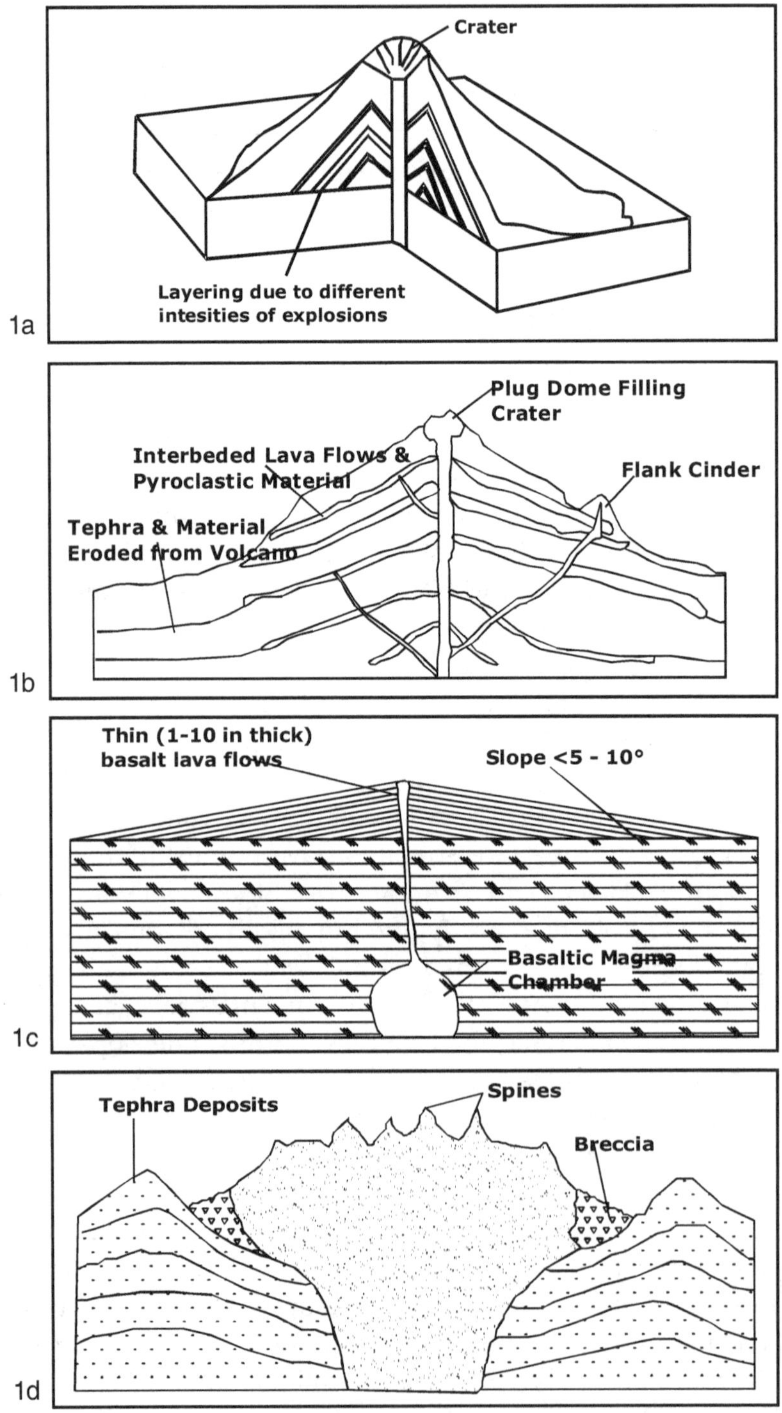

Figure 1. Schematic diagram depicting types of volcanoes

Classification based on Activeness

Based on "recentness" of explosion, volcanoes are termed as "active" or "dormant" or "extinct". However, as the lifespan of a volcano may vary from few days to many millions if not many tens of millions of years, these are relative terms and the meaning of which vary regionally. As of today, there is no consensus among volcanologists about these terminologies.

An "active volcano" denotes a volcano, which has erupted recently in the historic past and shows signs of activeness in terms of gaseous emission, seismic activity and abnormal thermal anomaly in and around it. An "extinct volcano" denotes a volcano whose vent has been solidified upto the volcanic chamber (?) and shows no seismic activity and lacks thermal anomaly. The "dormant volcano" is a transitional variety between these two extremes but shows no signs of emission in near future or at least has not shown any such signs in recent history. However, given cognizance to the dynamic plate movements and continuously changing earth's surface, there can be only two types of volcanoes - active and dormant.

The Table 1 shows major volcanoes and their eruption events as recorded in historic perspective and recent documentations (*Source:* Keller, 2005).

Classification based on Nature of Eruption and Material

An eruption dominated by the outpouring of lava onto the ground is often referred to as an "effusive eruption" (as opposed to the violent fragmentation of magma by explosive eruptions). Lava flows generated by effusive eruptions vary in shape, thickness, length and width depending on the type of lava erupted, discharge, slope of the ground over which the lava travels and duration of eruption. "Phreatic eruptions" are steam-driven explosions that occur when water beneath the ground or on the surface is heated by magma, lava, hot rocks, or new volcanic deposits (for example, tephra and pyroclastic-flow deposits). The intense heat of such material (as high as 1,170° C for basaltic lava) may cause water to boil and flash to steam, thereby generating an explosion of steam, water, ash, blocks, and bombs. "Plinian eruptions" are large explosive events that form mammoth dark columns of tephra and gas high into the stratosphere (>11 km). "Strombolian eruption" is characterized by the persistent ejection of lava fragments and scoriae, stromboli. This type of eruption also undergoes periods of rhythmic vapor emission and violent eruptions.

Each episode is caused by the release of volcanic gases, and they typically occur every few minutes and the lava fragments generally consist of partially molten volcanic bombs that become rounded as they fly through the air. A "vulcanian eruption" is a type of explosive eruption that ejects new lava fragments that do not take on a rounded shape during their flight through the air. This may be because the lava is too viscous or already solidified.

Table 1. Historical volcanic events, associated damage and eruption type

Volcano or City	Year	Effects
Vesuvius, Italy	A/D. 79	Destroyed Pompeii and Killed 16,000 people. City was buried by volcanic activity and rediscovered in 1596.
Skaptar Jokull, Iceland	1783	Killed 10,000 people (many died from famine) and most of the island's livestock. Destruction felt up to Scotland.
Tambora, Indonesia	1815	Global cooling; produced "year without a summer."
Krakatoa, Indonesia	1883	Tremendous explosion; 36,000 deaths from resultant tsunami.
Mount Pelee, Martinique	1902	Ash flow killed 30,000 people in a matter of minutes.
La Soufriere, St.Vincent	1902	Killed 2000 people and caused the extinction of the Carib Indians.
Mount Lamington, Papua New Guinea	1951	Killed 6000 people
Villarica, Chile	1963-64	Forced 30,000 people to evacuate their homes
Mount Helgafell, Iceland	1973	Forced 5200 people to evacuate their homes.
Mount St. Helens, Washington, USA	1980	Debris avalanche, lateral blast, and mudflows killed 54 people, destroyed more than 100 homes.
Nevado del Ruiz, Colombia	1985	Eruption generated mudflows that killed at least 22,000 people.
Mount Unzen, Japan	1991	Ash flows and other activity killed 41 people and burned more than 125 homes. More than 10,000 people evacuated.
Mount Pinatubo, Philippines	1991	Tremendous explosions, ash flows, and mudflows combined with a typhoon killed >300 people; 1000's of people evacuated.
Montserrat, Caribbean	1995	Explosive eruptions, pyroclastic flow; south side of island evacuated, including capital city of Plymouth; several hundred homes destroyed.

Moderate explosive eruptions commonly eject a large proportion of volcanic ash and also bombs and blocks of crustal materials. Andesitic and dacitic magmas are most often associated with vulcanian eruptions, because their high viscosity (resistance to flow) makes it difficult for the dissolved volcanic gases to escape except under extreme pressure, leading to explosive behavior. Simkin *et al.* (1981) summarized data on eruptive characteristics. According to them, of the 5564 dated eruptions recorded, 24.1% included lava flows, 8.8% were of phreatomagmatic type, 5.1% produced pyroclastic flows, 5.6% generated mudflows or lahars and 5.9% were submarine eruptions. An overwhelming 72.8% of the dated eruptions involved Strombolian, Vulcanian or Plinian explosive activity.

OCCURRENCE

More than 10,000 volcanoes have been mapped all over the earth, among which, nearly 500 are considered to be active, out of which, more than half are in the Pacific Ocean forming a circle, called Pacific Ring of Fire. Indonesia in the Pacific Ocean has the most number of active volcanoes. Since the early 1800's, there have been over 600 eruptions from Indonesia's 70 volcanoes. Central America has 60 active volcanoes, the Soviet Union 50, Japan 45, Africa 40, Iceland 20, and Italy 12. There are also active volcanoes in the Mediterranean Sea, across the southern part of Europe, and on the floor of the Atlantic Ocean.

Principally, most number of volcanoes occur along divergent plate boundaries, followed by continental extension areas and finally along converging plate boundaries. Volcanoes of divergent plate margins are located in the Mid-Atlantic Ridge located across Iceland, East African Rift Valley, Rift valley across Red Sea, Oceanic ridge of Gulf of Aden Basin and Range Province of the western U.S.A. and Rio Grande Rift, New Mexico. Pacific Ring of Fire is an example of converging plate margin. The convergent boundaries along the coasts of South America, Central America, Mexico, the northwestern U.S.A., western Canada and eastern Alaska are boundaries in which oceanic lithosphere is being subducted beneath continental lithosphere. This has resulted in the formation of continental volcanic arcs that form the Andes Mountains, the Central American Volcanic Belt, the Mexican Volcanic Belt, the Cascade Range, and the Alaskan volcanic arc. The Aleutian Islands (west of Alaska), the Kurile-Kamchatka Arc, Japan, Philippine Islands, and Marianas Islands, New Zealand and the

Indonesian Islands, along the northern and western margins of the Pacific Ocean are zones where oceanic lithosphere is being subducted beneath oceanic lithosphere. These are all island arcs.

Volcanism also occurs in areas that are not associated with plate boundaries, in the interior of plates. These appear due to upwelling of hot mantle material towards the surface, independent of the convection cells. Hot spots tend to be fixed in position, with the plates moving over the top. As the rising plume of hot mantle moves upward it begins to melt to produce magmas. These magmas then rise to the surface producing a volcano. But, as the plate carrying the volcano moves away from the position over the hot spot, volcanism ceases and new volcano forms in the position over the hot spot. Sequence of this process produces chains of volcanoes or seamounts (former volcanic islands that are eroded below sea level). Volcanisms resulting from hot spots occur in both the Atlantic and Pacific Ocean, but are more evident on the sea floor of the Pacific Ocean, because the plates here move at higher velocity than those under the Atlantic Ocean. A hot spot trace shows up as a linear chain of islands and seamounts as can be seen in the Pacific Ocean. The Hawaiian Ridge is one such hot spot trace. Here, the Big Island of Hawaii is currently over the hot spot, the other Hawaiian Islands still stand above sea level, but volcanism has ceased. Northwest of the Hawaiian Islands, the volcanoes were eroded and are now seamounts.

INDIAN VOLCANOES

India has one of the largest volcanic provinces on earth - the Deccan Plateau. The lava flows emanated from Deccan volcanism cover an area of about 900 square kilometers and the volcanism lasted for about 2 to 3.5 million years. Fortunately, the Deccan volcanism is an event of past that lasted during 67 and 65 million years ago and was formed by the Reunion hotspot. Deccan volcanism coincided with the decline of the dinosaurs raising the possibility that the Indian volcanoes were involved with their decline. However, recent researches have indicated that the impact of Deccan volcanism could not have made significant impact even in southern India and hence global impact of this event towards extinction of dinosaurs is questionable (Ramkumar *et al.* 2004; 2005). Notwithstanding these debates, the quantum of lava flow, possible amounts of CO_2 ejection, duration of volcanism and the area covered by lava are all unparalleled in earth's history. Furthermore, the volcanic event is a thing of geologic past and poses no activity since 65.4 million years before present.

Recent volcanic activity in Indian Territory includes Barren Island volcano, Narcondum volcano in Andaman Islands and an unnamed volcano located under submarine conditions. Barren Island volcano covering an area of about 10 square kilometers and rising to about 335 m above mean sea level is an active volcano since historic past. Documented evidence indicates that it erupted during the year 1787 and since then shown volcanism in the years of 1789, 1795, 1803-04, 1952, 1991, 1994-1995, 2000 (Raina 1987; Haldar 1989; Haldar et al. 1992, 1995; Ravi *et al.* 2001; Haldar and Luhr 2003). Most recently, this volcano had erupted on 28th December 2004 and it was followed by a series of eruptions in January, May and October 2005 with greater intensities (Bandopadhyay *et al.* 2006). The Narcondum volcano is located in Andaman Islands, India. It covers an area of 12sq.km, with a summit elevation of about 710m above MSL. It is a strato volcano, erupting andesitic lava. It was previously recorded as extinct. After being dormant for about 100 years, during June 2005, this volcano started ejecting mud and smoke since Sumatra earthquake of the year 2004. An unnamed submarine volcano located at 156km SE of Chennai and 100km east of Pondicherry is said to have erupted (John Seach 1757) during the year 1757. Its existence is yet to be confirmed.

FORECASTING

Like many other geohazards, volcanic hazard could also be not predicted accurately. However, unlike other geohazards, there are many other indirect methods of observations and indications that could forecast probable volcanic activity in a particular region of interest. Sophistications in instrumentation and sensors in the recent years led to ascertaining the probability of volcanic activity. Over the years, scientific community could develop few generalized methods in forecasting volcanic activity. These methods include observation of seismic waves and their pattern in places of interest, detection of gaseous emissions in terms of presence/absence and quantum of volcanic gases, detection of sudden appearance of thermal anomalies in water bodies, change in magnetic anomalies of an area, slow disappearance of vegetation, abnormal behavior of animals, deformation of ground surfaces and sudden appearance of glassy materials in strange and unexpected places. Furthermore, the remote sensing satellites send imageries of earth features routinely at fixed intervals of time duration. Comparison of these imageries for any significant change in regions in and around known volcanic spots has become a regular exercise for detecting volcanism.

MITIGATION MEASURES

The countries with the highest volcanic risk are those which contain active volcanoes, such as Italy, Indonesia, New Zealand, Papua New Guinea, Philippines, Japan, Russia, United States of America, Mexico, Central America, Colombia, Equator, Peru and Chile. From the past experiences of volcanism of these countries, following mitigation measures are suggested.

- Foremost mitigation measure is to avoid regions of volcanic hazards from dwelling, establishment of infrastructure and other structures. Only few volcanoes exist in India and they are not inhabited places and hence volcanic hazard is abysmal in India.
- It is said that volcanic eruptions cannot be prevented. However, although the volcanic explosion is sudden, the time required for simmering, uprising of lava from deep inside of earth to near surface, time lag in building up of pressure before explosion have all invariably show many indications on physical properties of surface of earth besides change in atmospheric chemical composition, that may provide adequate time to evacuate and initiate safety measures. Thus, once a volcano is located, continuous monitoring of the region could provide enough advance notice to make arrangements for evacuation and precautionary measures.
- Even in thickly inhabited regions that are located adjacent to active volcano, there have been moderately successful attempts of damming and diversion of lava/mud/ash flow away from infrastructure and establishments. While this measure is applicable only to effusive type of volcanism, the delivery of volcanic ash, lava, pyroclastic materials and other gaseous materials by explosive type of eruption could not be contained.
- There have been attempts to bomb and diffuse volcanic explosion with fairly better results. Draining of volcanic crater lakes and spraying water over lava flows have also been attempted to save lives and infrastructure.

References

Bandopadhyay, P C., Sumit, K M. and Tapanpal. T., 2006 The 2005 Eruption on Barren Island, Andaman Sea. Curr.Sci. v.90. pp.620-622.

Haldar, D., 1989 Petrology and chemistry of recent volcanism of Barren and Narcondam islands - The only two recent volcanoes in India. Rec.Geol.Surv.India. v.122. pp.48-49.

Haldar, D., Lasker, T., Bandyopadhyay, P.C., Sarkar, N.K. and Biswas, J.K., 1992 The volcanic eruption of the Barren Island volcano, Andaman Sea. Jour.Geol.Soc.India. v.39. pp.411-419.

Haldar, D., Chakraborty, S C. and Chakraborty, P P., 1995 The 1995-eruption of the Barren Island volcano in the Andaman Sea. Rec.Geol.Sur.India. v.129. pp.59-61.

Haldar, D. and Luhr, J F., 2003 The Barren Island volcanism during 1991 and 1994-95: eruptive style and lava petrology. Mem.Geol.Soc.India. v.52. pp.313-338.

John Seach., 1757 The world's leading volcano Adventures, New South Wales, Australia.

Keller, E.A., 2005 Introduction to Environmental Geology, Prentice Hall, 3rd Edition. 672p.

Raina, V K., 1987 A note on sulphur occurrence in the volcanoes of the Bay of Bengal. Ind.Miner. v.41. pp.79-86.

Ravi, S., Haldar, D., Absar, A. and Chakraborty, S.C., 2001 Pictorial monograph of Barren Island volcano: The lone active volcano in the Indian Subcontinent. Geol.Surv.India.Spl.Pub. v.67. pp.84-88.

Ramkumar, M., Stueben, D., Berner, Z. and Schneider, J., 2004 Geochemical and isotopic anomalies preceding K/T boundary in the Cauvery basin, South India: Implications for the end Cretaceous events. Curr.Sci. v.87. pp.1738-1747.

Ramkumar, M., Harting, M. and Stüeben, D., 2005 Barium anomaly preceding K/T boundary: Plausible causes and implications on end Cretaceous events of K/T sections in Cauvery basin (India), Israel, NE-Mexico and Guatemala. Inter.Jour.Earth.Sci. v.94. pp.475-489.

Sathish, G., 2008 Earthquake-The creator of geohazards. *In:* Ramkumar, M., (Ed.) Geohazards: Causes, consequences and containment. New India Publishers, New Delhi. pp.(in press).

Simkin, T., Siebert, L., McClelland, L., Bridge, D., Newhall, C. and Latter, J.H., 1981 Volcanoes of the World. Hutchinson-Ross, Stroudsburg, Pennsylvania. 232p.

❑❑❑

9

Characterization and Mitigation of Landslides in the Nilgiri Hills, Tamilnadu

G. MANIMARAN
School of Tectonics, Department of Geology, V.O.C. College, Tuticorin - 628 008.
E-mail: <acugemmani@yahoo.co.in>

Readers are introduced through this chapter to a geohazard (Landslide) that is most frequently publicized in mainstream media. Systematic examination of landslides and associated structural, tectonic, geological, geomorphological, meteorological, hydrogeological characteristics and land use pattern of Nilgiri hill area had indicated that steep slope, dipping of rocks towards slope, presence of clay layering in weathered rock types, events of continuous heavy rains, flooding on the slopes and improper land use practices play a major role in the genesis of landslides. In addition, ongoing seismotectonic activities in Southern Granulite Terrain of South India and reactivation of hidden blind shear zones were identified to be causative factors of recent increase of landslide occurrences.

INTRODUCTION

The downward movement of consolidated and unconsolidated soils and rock *en masse* from any geomorphic feature due to natural or manmade causes under the influence of gravity is termed as landslide (GSI, 1982). Nevertheless, presence of water, clay material, weak joints, favorable slope and many other characteristics also act as catalysis for this event. Landslides are one of the common geological hazards that affect countless population in India, especially in hilly terrains. Different reasons have been ascribed to different landslides of various terrains. Among them, reactivation of Bhavani shear zone (Prasannakumar and Llyod, 2007), high anthropo-geomorphic intervention (Basu, 2005), local geology, precipitation and slope (Bhattacharjee, 2005), presence of local vertical faults, triangular facets, abrupt deflection of streams and neotectonism (Singh, 2007) are considered to be classic examples pointing the multi-causal genesis of landslides. Using Geoenvironmental factors like lithology, slope, morphometry, structure, relative relief, land use and land cover and hydro-geological conditions, Naithani (2007) has developed a rapid hazard assessment technique and proposed Landslide Hazard Evaluation Factor (LHEF). This chapter attempts to characterize the landslide phenomenon and based on which suggests mitigation and safe guarding measures.

GEOLOGY OF NILIGIRI HILLS

Nilgiri Hills of Tamilnadu are located at the junction of the Eastern and Western Ghats. The country rock exposed here is charnockite (granulite). The trident shaped Nilgiri granulite terrain of exhumed nature extends for a strike length of about 140 km with a maximum width of about 80 km. Two major shear zones namely, Moyar shear zone (MSZ) and the Bhavani Shear Zone (MBSZ) border the Northern and Southeastern margins of Nilgiri granulite terrain respectively (Fig.1). General strike of the granulites is N60-70°E with steep dips (55°-75°). The charnockities are syn-accretionary granulites that were accreted to the Dharwar craton prior to the 2500 Ma granulite event (Janardhan et al. 1994). The original layering structures are well preserved in charnockite. The dark colored melanocratic layer of these granulites has garnet and biotite alternating with quartz and plagioclase rich layer. This layering was attributed to extensive migmatization followed by granulite high-grade metamorphism (Srikantappa, 1996). On alteration, the charnockites get converted into thick brownish weathered skin with clay layers formed from feldspars. Apart from charnockites, there are bands of gabbroic anorthosites, anorthosites, mafic granulites, pyroxenites and banded magnetite quartzites and dykes of dolerites. Topsoil of the study area is represented by residual, lateritic red coloured porous soil. Buried lignite deposits are also seen in valley portions. Black soil of swelling nature, enriched in montmorillonite and kaolin are also present as top regolith in some places, where basic and ultrabasic rock types occur.

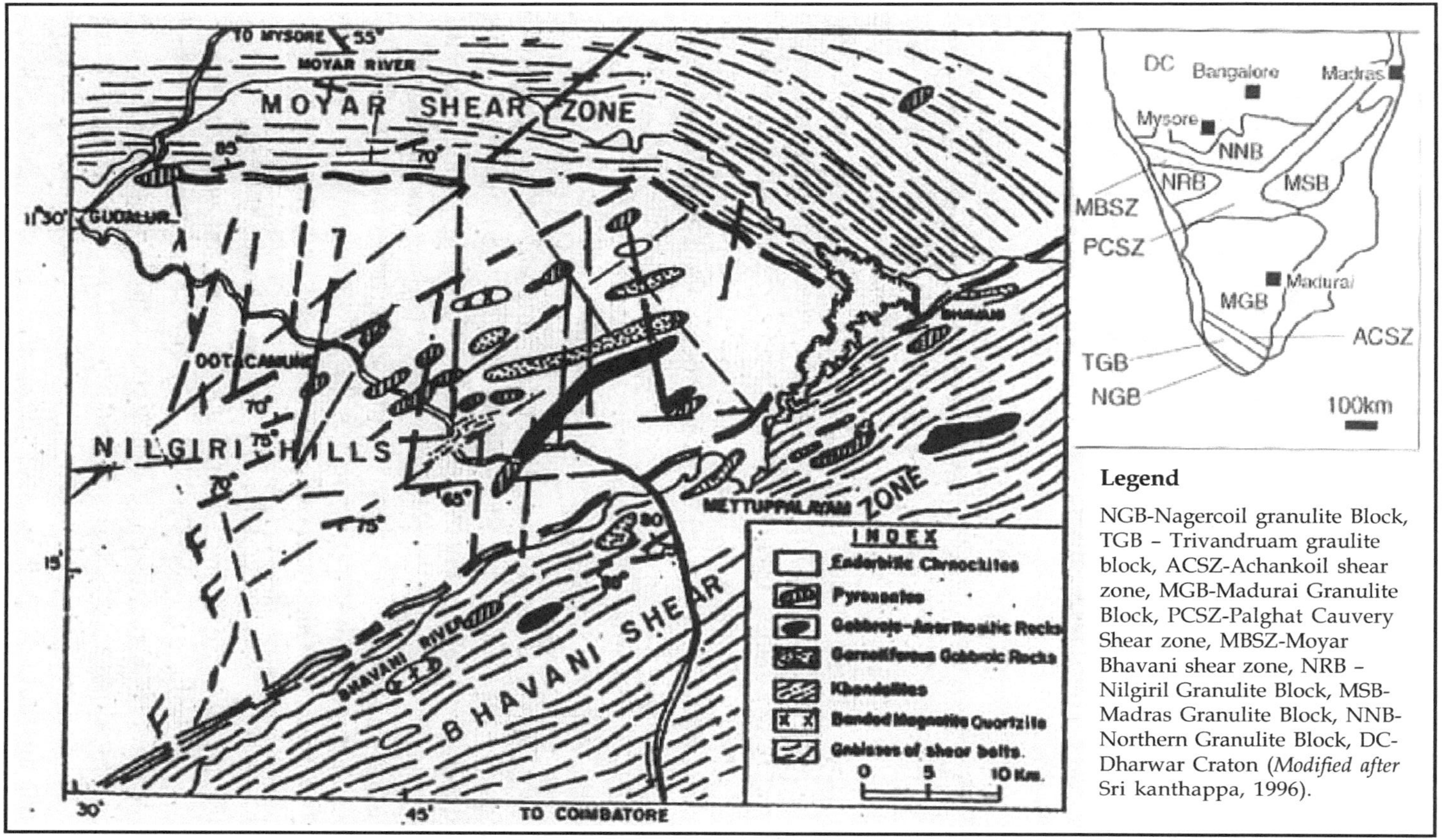

Figure 1. Location Map of the Study area showing lithology, major structural trends and shear zones.

LANDSLIDES IN NILGIRI HILLS AND ASSOCIATED TRAITS

Occurrence of landslide was rare in the history of the Nilgiris, but of late, it has become an annual phenomenon, causing havoc during the northeast monsoons. Landslides were severe during the two consecutive years 1978 and 1979 as well as in 2006, inflicting loss of life and damage to property. Unprecedented rains have triggered about 100 landslides within an area of 250sq.km during 1978 while nearly 200 and 30 landslides were recorded during 1979 and 2006 respectively. In addition, subsidence, soil and rock slips, earth slides (Photograph 1) and rock falls became common immediately after every major rain.

The area in which beds dip towards the slope with dip amounts more than 45^0 are unsafe (Arora, 1988) and prone to landslides (Fig. 2). It is observed in many locations of landslide occurrences in Nilgiri hills that the rock formations are invariably dipping towards topographic slope. It is also observed that whereever beds are dipping into the hills and against the slope (Fig. 3), even rock excavations for road and railway lines are comparatively free of landslides. Improper land use practices such as heavy tilling, agricultural practices and settlement patterns and withdrawal of toe support have contributed to creep in many cases. A common factor noticed in most of these vulnerable slopes are deforestation in the recent past, cultivation of seasonal crops and increase in settlements.

It is observed that while there were innumerable slides along road cuttings between Coonoor and Mettupalayam, the number of slides on the Kotagiri-Mettupalayam sector was negligible. Along this road, the natural forests remain relatively undisturbed and topographic slopes are moderate to steep. In the Kotagiri Section, geological formations are mostly dipping against (or) oblique to the geomorphic slopes of area and thickness of weathered cover is low.

It is commonly observed that the charnockites are intensely jointed in multiple directions, which in turn have paved way for severe weathering, making the terrain weak in nature. Furthermore, the joints act as conduits of water seepage. When such joints are inclined towards topographic slope and the terrain is cut for road, railway or agricultural or urban and or any other infrastructure development, landslide becomes inevitable.

Photograph 1. Landslide in road section between Mettupalayam and Marapalam.

Figure 2. Diagramatic representation of Nilgiri Road section between Mettupalayam and Coonoor shows steep beds dipping towards the slope - unsafe condition.

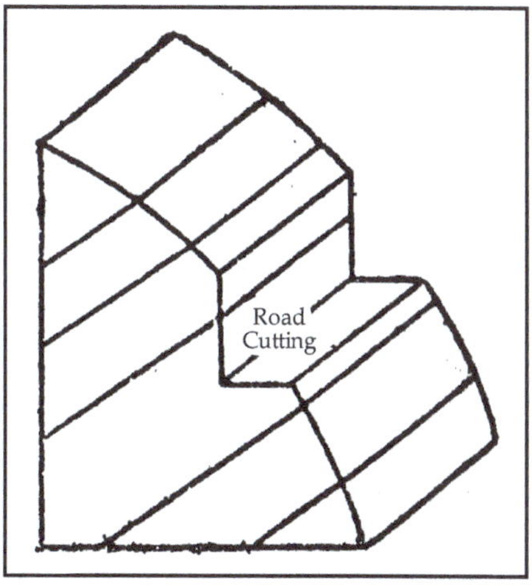

Figure 3. Kothagiri to Mettupalaym Ideal road section showing beds dipping into the hill slope -safe condition.

Photograph 2. Highly jointed and weathered charnockite. Location of the photograph is near Kundah dam.

Nilgiri hill slopes are mostly covered with either red colored lateritic soil or black colored soil. As these residual soils occur over steeply dipping, highly weathered and jointed country rock, any alteration of natural conditions including deforestation triggers landslide.

The lateritic soils are intercalated with thin layers of kaolinite clays and black soils are enriched with montmorillonite clays. During rainy season, the clay beds act as smooth gliding surfaces and facilitate landslide. Bulging and swelling of clays during wet season followed by shrinkage of clays during consecutive dry season result in a subsidence of ground. Clay bed present along the fracture surface of rock formation if wet acts as a lubricant and greatly enhances the possibilities of rock slips (Fig.4). Vibrations created by heavy vehicular traffic may also accelerate the landslides especially during rainy season and flooding on the slopes.

In the Nilgiri Hills, it has been found that steep as well as gentle slopes have failed. Comparison of steepness of slope and the number of landslides indicate that the slides in 1978 and 2006 occurred on comparatively steeper slopes than those of 1979. This correlation, together with rainfall data reveal that the duration of precipitation in 1978 and 2006 was high only for three days, whereas the rainfall in 1979 spanned for about more than two months. This observation indicates that the 1978 and 2006 slides occurred due to heavy precipitation in a short period when there were flash floods and consequent soaking on the slopes resulting in mass movement of the material over relatively steeper slopes. On the contrary, during the 1979 monsoon, the longer duration of rainy period permitted greater infiltration of water into the soil and consequent triggering of landslides in areas as gentle as 10^0 slopes. Slides were observed in areas of anthropogenic activity including agricultural operations, construction of buildings and road, removal of earth and rock, leveling of slopes, deforestation and blocking of natural drainages. In addition, slopes have failed where the toe had been removed either by stream action or by human interference. Deforestation has marked effect in rendering the slopes slide prone. In the 1978 landslides, deforested slopes have failed while the adjacent areas with similar topography and geology remained unaffected possibly because of vegetal cover (Raja, 2008).

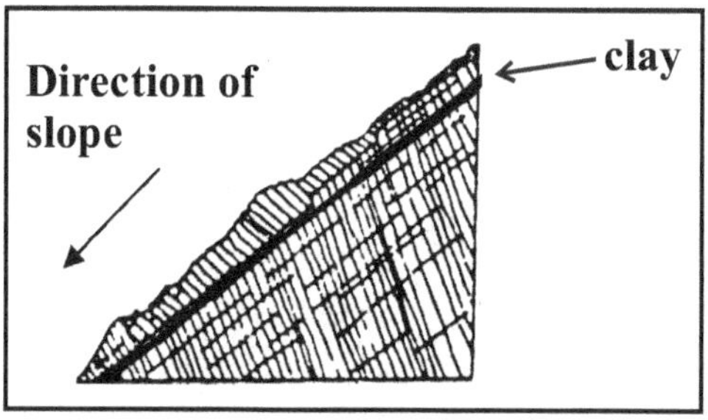

Figure 4. When rainwater enters soils with clay layer it acts as a lubricant triggering sliding of overlying rock unit.

Kilometer scale compound slides are observed at Katteri hill site (Photograph 3). Both soil and rocky materials were moved from the top of the hills to the bottom. The entire slope was subjected to downward movement, which suggests that the slide might have happened along a hidden blind shear zone preserved beneath the lateritic soil cover. It is possible that many landslide sites including Glendale Tea Estate, Elk Hills, Allangy, Burliar and Yelanathi may represent the locations of paleo shearzones.

Tectonically, the Nilgiri granulites are squeezed between E-W Moyar shear zone in the North that moves dextrally and NE-SW trending Bhavani Shear zone that moves in the Southeast sinisterly. Presence of these major sheer zones along the boundaries of the Nilgiri hills and the presence of landslide prone blind sheer zones suggest geodynamically active nature of the major sheer zones. This inference is supported by the observations of Chattopadya et al. (2006) who have documented upliftment characters and reactivation of lineaments of Western Ghats.

Photograph 3. Large scale Katteri compound slide site showing slided large slope along a narrow line- a blind shear zone.

There is a noticeable increase of landslides in Nilgiri hills especially after Republic day earthquake on 26.1.2001 in Gujarat. Far-field foreshock of the 2001 Gujarat earthquake was realized at Nanguneri of Tamilnadu on January 17, 2001. Since then, the Southern Granulite terrain of Tamilnadu and Kerala have been experiencing many microseismic fracturing events of lands and buildings usually accompanied with either seismic noise or vibrations of household utensils (Manimaran and Renuga, 2007). These microseismic fracturing events reveal the ongoing reactivation of crustal segments bounded by N-S, NW-SE, NE-SW and E-W trending deep shear zones of South India (Manimaran 2007; Manimaran and Renuga, 2007), which in turn could also have contributed towards increase of landslide events in the Nilgiri hills.

RECOMMENDATIONS AND MITIGATION MEASURES

From the foregoing observations and interpretations, it is apparent that the Nilgiri hill region has become active landslide prone area, due to many reasons including natural geotectonic activation, deforestation, unscrupulous developmental activities, etc. At this juncture, it is inevitable to be prepared and adopt measures that could reduce the impacts of landslides. In view of this, measures suitable for different developmental activities are enlisted (Ahamed and Gopikrishnan, 2008; Manimaran, 2008).

General

- Slopes are very steep at Nilgiri. Hence excavations at the base of the slopes should not be allowed.
- The entire region should be mapped and different thematic maps detailing natural geological conditions, soil properties, slope, etc, have to be prepared and local governmental agencies should take adequate precautions before sanctioning any developmental activity that may alter natural conditions.
- In slide prone areas, due restrictions on heavy traffic, building construction and use of explosives (blasting) should be imposed.
- Slope, if very steep, may be strengthened through growing natural vegetation, grouting, geotextiles, etc.
- Dumping of extra material at the toe (base) of the slope provides more stability.
- If the shearing stress is too high, material from the head (top) of the slope may be removed.
- Retaining walls, wherever possible, should be raised to hold back the unsupported rock masses. Anchoring may strengthen these walls.
- Construction of new building in the sites where rock formations dipping along slopes of the hills is to be banned.

- Seepage of water must be checked, so that the slope material remains dry. The surface drainage should be planned in a very appropriate way so that water does not accumulate anywhere.
- Local conditions have to be examined and appropriate provisions such as drainage trenches, deep drainage, mulching, geotextiles, Gabion meshes, coir and geojute netting and bench terracing have to be made.
- Turfing and forestation may stabilize loose soils on the slope.

Railways

- Periodic and frequent desiltation of the entire toe, side and cross drains and water collection pits all along the railway lines, have to be conducted.
- Systematic study by a research team comprising geologists and geoengineers has to be conducted. Landslide preventive measures suggested by their recommendations have to be should be completed as early as possible.

Roads

- Chute, vertical drains and parapet walls must be provided with roadside retaining walls wherever necessary.
- All culverts, junction boxes and side drains should be desilted periodically and more frequently during monsoon.
- Drinking water pipes and telephone cables should be laid only on the surface of the earth.
- Leak proof sewage and rainwater channels should be constructed for drainage purposes along urban roads and they should be connected with the nearby existing natural drainages.
- The waterway of bridges must be cleared of vegetation, silt and other obstructions.
- During desilting, care should be taken to remove the silt only upto bed level.
- Water shall not be allowed to stagnate on the surface. For this purpose, the entire toe drains and sewage drains should be cleaned and maintained without blockage.

- Loose boulders should be cleared in a systematic manner all along the road cuttings in the hill slopes.
- Alternative road system has to be established for heavy vehicle movement.
- Advance blasting methods have to be followed to remove the loose materials and boulders in a systematic manner.

Electricity and Telecommunication

- All electric and telephone poles including high-risk towers and transformers should be erected with strong RCC foundations and the strength of pole concrete should be checked before monsoon and periodically during monsoon.
- Site selection for installing CDMA tower should be made by competent geologists and geoengineers.

Buildings

- Building permits have to be granted only in regions safe from landslide hazard. Strict vigil has to be maintained for any violation.
- Buildings have to be constructed with appropriate provisions according to safety measures.
- Pre monsoon checking of foundation, surfaces and drains, etc. have to be conducted by competent scientific personnel in association with local authorities.
- Water dampness should not be allowed around the buildings.
- Special vigil has to be kept over buildings situated on or near steep slopes.
- Blockage of surface and sub surface drainage should be avoided. Periodic checking and cleaning should be conducted.
- Angular retaining walls with weep holes should be constructed to protect the angular earth cut faces around the buildings.
- All buildings should be provided with damp proof course at the basement level.
- Leak proof sewage and drainage channels should be constructed to avoid infiltration.

- Vacant land portions around the houses have to be protected by cement grouting or plastering as per the local conditions without affecting natural percolation.
- Man made soil erosion activities should not be permitted in and around the residential areas.
- Sufficient distance should be provided between the cut face and the buildings.
- Rainwater harvesting sump should be placed on the surface of the earth.
- Planting shallow rooted plants and turfing should be done in the residential areas.

CONCLUSIONS

The pattern and nature of landslide occurrence in the Nilgiri hills are examined and found that landslides occur frequently during monsoons. Irrespective of vegetated or non-vegetated, steep or gentle slopes, landslides occur everywhere, particularly in regions wherein natural geological, geomorphological and drainage conditions were altered. It is suggested that a systematic microscale study of the entire hill region is made and landslide hazard zonation map is prepared. Any developmental activity in this region should be based on the results and recommendations of the study in order to minimize the potential impact of landslide hazard. Although few studies have been conducted in this direction, the scales were regional in nature and hence microscale study is suggested with a view to enlist site and specific mitigation measures, which in turn, on implementation, could potentially reduce the landslide hazard in the Nilgiri hills.

References

Ahamed, S.B., and Gopikrishnan, S., 2008 Genesis and control measures for landslides in the Nilgris. *In:* Singh, D.V., Selvi, V. and Khola, O.P.S., (Eds.) Workshop on Landslides in the Nilgris, 5th March, 2008- Compedium of Lead papers, CSWCRTI Research Centre, Udhagamandalam, Tamil Nadu. pp. 23-37.

Arora, D.S., 1988 A Textbook of Geology, Capital Publishers, Chandigarh, 444p.

Basu, S.R., 2005 Quantitative assessment of Anthropo-Geomorphic processes in Darjiling town. *In:* Chandrasekar, N. (Ed.) Quarternary Climatic changes and landforms. Manonmaniam Sundaranar University, Tirunelveli. pp.159-172.

Bhattacharjee, D., 2005 Natural hazards of Sikkim Himalayan in with Special reference to landslides hazards. *In:* Chandrasekar, N. (Ed.) Quarternary Climatic changes and landforms. Manonmaniam Sundaranar University, Tirunelveli. pp.173-188.

Chattopadhyay, S., Sajikumar, S and Chattopadhyay, M., 2006 Landslide evolution in parts of Vamanapuram Drainage Basin, Kerala - A Hypsometric Approach. Jour.Geol.Soc.India. v.68. pp.841-856.

Geological Survey of India., 1982 The Nilgiri Landslides. Misc.Publ.No. 57.

Janardhan, A.S. Jayananda, M. and Shankara, M.A., 1994 Formation and tectonic evolution of granulites from the Biligiri and Niligiri hills, S.India: Geochemical and isotopic constrains. Jour.Geol.Soc.India. v.44. pp. 27-40.

Manimaran, G., 2007 Indian Ocean Tsunami and related events. Renuga Publications, Tirunelveli. 72p.

Manimaran, G. and Renuga, K., 2007 Understand and conquer the seismic earth (Tamil) Renuga Publications, Tirunelveli. 250p.

Manimaran, G., 2008 Genesis and Characterization of Landslides in the Nilgris. *In:* Singh, D.V., Selvi, V. and Khola, O.P.S., (Eds.) Workshop on Landslides in the Nilgris, 5th March, 2008- Compedium of Lead papers, CSWCRTI Research Centre Udhagamandalam, Tamil Nadu. pp.1-11.

Naithani, A.K., 2007 Macro Landslide Hazard Zonation Mapping using univariate statistical analysis in a part of Garhwal Himalaya. Jour.Geo.Soc.India. v.70. pp.353-368.

Prasannkumar, V. and Lloyd, G.E., 2007 Development of crystallographic lattice preferred orientation and seismic properties in Bhavani Shear Zone, South India. Jour.Geo.Soc. India. v.70. pp 282-296.

Raja, R., 2008 Landslide control measures in Nilgris. *In:* Singh, D.V., Selvi, V. and Khola, O.P.S., (Eds.) Workshop on Landslides in the Nilgris, 5th March, 2008- Compedium of Lead papers, CSWCRTI Research Centre Udhagamandalam, Tamil Nadu. pp. 38-47.

Singh, T., 2007 Geology of Itanagar Captial Complex, Arunachal Himalaya, with special reference to neotectonics. Jour.Geo.Soc.India. v.70. pp.339-352.

Srikantappa, C., 1996 The Nilgiri Granulites. *In:* Santosh, M. and Yoshida, M. (Eds.) The Archaean and Proterzoic terrains in Southern India within East Gondwana.pp.185-222.

❑❑❑

10

Cyclone Disaster Management with Special Reference to Orissa Coast, India

DEVANANDA BEURA
Post Graduate Department of Geology, Utkal University, Bhubaneswar.
E-mail: <debanandabeura@rediffmail.com>

This chapter provides an introduction to cyclogenesis, nature of cyclone disaster and historic events. The mechanism that is in practice in cyclone mitigation, information dissemination and relief and rescue operations are detailed. As the state of Orissa is being battered more frequently by cyclones than any other state of India, special emphasis on cyclone disaster management system of Orissa is also discussed.

INTRODUCTION

Orissa state with immense natural resources is located between 17°27′N and 22°34′N latitudes and meridians of 81°27′E and 87°29′E longitudes. Orissa lies on the eastern coast of India with the waters of the Bay of Bengal swirling along its eastern and southeastern boundaries. It shares its boundaries with West Bengal and Jharkhand on the north, Chattisgarh on the west and Andhra Pradesh on the south. Out of the Indian mainland coastline of 7500 km, it constitutes a coastline of about 480 km. It extends for about 155,707 sq.km accounting about 4.74% of the total area of India. According to the 2001 census, it has 3.57 percent of the total population of India. Administratively, it has 30 districts, 3 Revenue Divisions, 58 Sub-Divisions, 51,061 villages and 127 Urban Centers.

Based on physiography, Orissa can be divided into three broad regions - the coastal plains, the middle mountainous region and the plateaus, and rolling up lands. The coastal plains extend from the River Subarnarekha in the north to the River Rushikulya in the south. The coastal zone covers nine districts namely Ganjam, Puri, Jagatsinghpur, Kendrapara, Jajpur, Khurda, Cuttack, Bhadrak and Balasore. It is narrow in the north, widest in the middle, narrowest in the Chilika coast and broad in the south. The coastal belt of Orissa lacks hills excepting few hillocks and mounds in the northern and southern parts. It has extensive low lands adjacent to the sea, sloping gently towards the shoreline of the Bay of Bengal. The coastline consists of the lakes, lagoons, marshes, bays and islands and natural harbors. The lakes in Orissa are natural and artificial. One of the blazing example of the natural lake is the Chilika lake.There is only one bay in Orissa which is the Hukitola bay. Eventually there are only two islands in Orissa - the Short's Islands and the Wheeler Islands. Among the natural harbor, there is only one of its type in Orissa i.e. Paradeep which holds its recognition as the deepest natural port in the whole of India. The main drainage system of this area is represented by several rivers at their old stages flowing sluggishly towards the sea, which form topographic features like peneplains, natural levees, deltas, etc. (Fig.1). The major rivers flowing in the coastal belts are the Subarnarekha, the Budhabalanga, the Baitarani, the Brahmani, the Mahanadi and the Rushikulya. This region is the combination of several deltas of varied sizes and shapes formed by these rivers. The coastal belt has an area of 40,166 sq. km with a population of 68,79,379, which is 45 percent of the population of 9 coastal districts of Orissa.

The climate of Orissa is of sub-tropic to tropical nature. The southwestern and the northeastern monsoons primarily control its climate. The state receives a total rainfall of about 1500 mm of which, about 1200 mm is received within four months from June to September. During the subsequent period, the rainfall is uncertain. In recent years, a slight deviation from normal climatic cycles is observed.

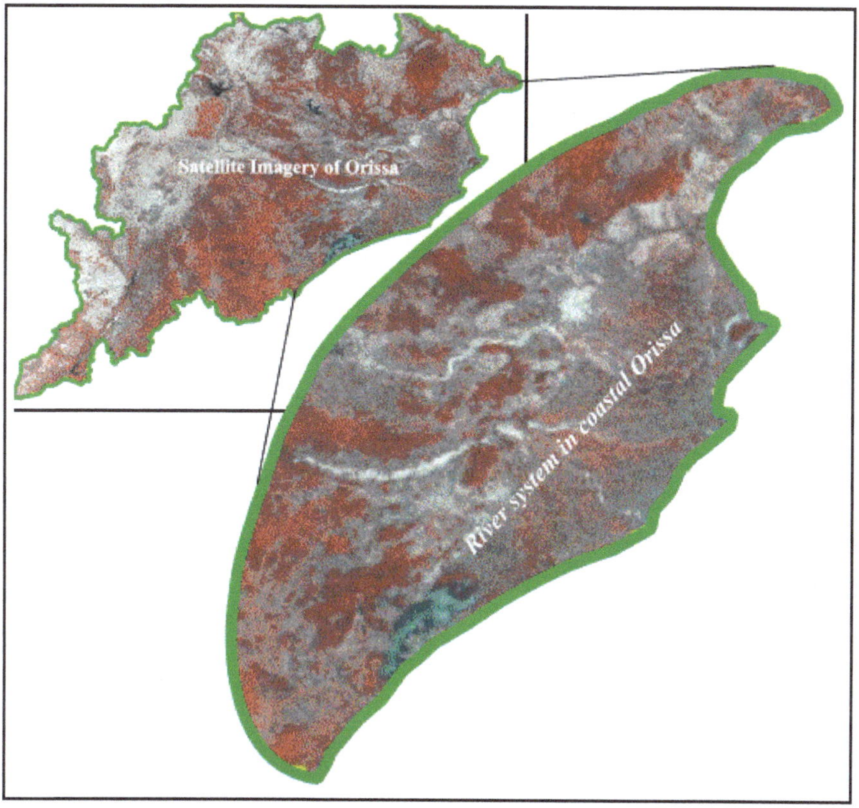

Figure 1. Satellite imagery of Orissa showing topographic features and rivers

COASTAL ORISSA - THE LAND OF DISASTERS

Within a span of 50 years from 1950 to 1999, Orissa was affected several times by natural calamities, thereby paralysing the social, economic and environmental conditions. The table 1 presents the natural calamities in the past 50 years. Geological and geophysical study shows that Orissa is not a highly prone to earthquakes as it is tectonically stable. The seismicity map indicates two earthquake risk zones in Orissa (BMTPC, 1997). The coastal area comprising the Brahmani Mahanadi graben and their deltaic areas come under Earthquake Risk Zone-III (Moderate Damage Risk Zone; Fig.2). Although many times tremors are felt, they are the resonances of earthquakes of distal regions. Heat stroke has emerged as the new natural disaster since the year 1998. It has become a regular phenomenon every year during April to June, as is the case of other cities of India.

Table 1. Occurrences of natural calamities in last 50 years on Orissa coast

Flood		Drought	Cyclone
1956	1982	1965	1967
1960	1985	1966	1968
1967	1990	1972	1971
1968	1997	1974	1978
1969	1999	1976	1980
1970	2001	1979	1982
1971	2003	1980	1986
1972	2007	1981	1987
1974		1986	1999
1975		1987	
1977		1988	
1980		1988	
1981		2000	

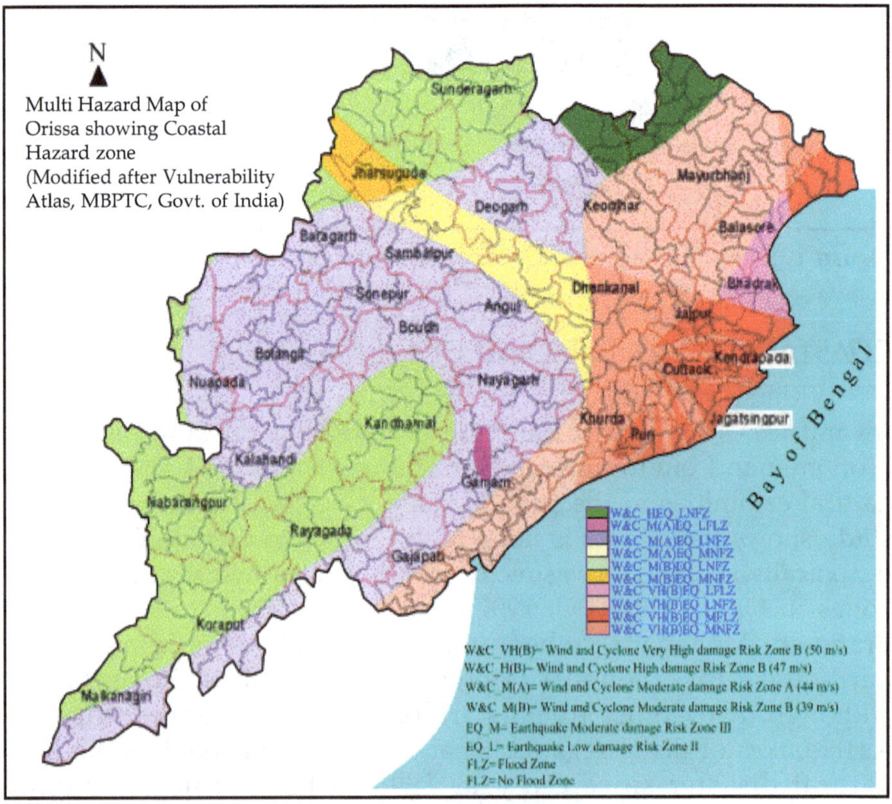

Figure 2. Multihazard map of Orissa
Source: Government of India

Orissa is considered as a state highly susceptible to multiple disasters (Fig. 2) and more particularly the coastal Orissa due to its sub-tropical littoral location is prone to tropical cyclones, storm surges and floods. It is densely populated along coastal plains that were creations of alluvial deposits of river systems. Sitting on these loose unconsolidated sediments are the developmental activities including habitats, commercial, industrial and recreational establishments in addition to strategic assets of defense forces. Whirlwinds and hailstorms usually appear in the months of March to June. In the pre-monsoon period (May and June) and post monsoon period (October and November), cyclonic disturbances occur in the Bay of Bengal. The coastal belt has experienced 1036 cyclonic depressions between 1891 and 1970. A total of 360 of these were intensified into Bay storms of which 221 were moderate and 139 developed into severe cyclones.

All the major rivers flow towards the lowlands of Orissa coast. The rivers in these areas have less carrying capacity but the heavy load of sediments introduced into these rivers results in frequent floods. The June to September are the likely months of flood. Out of total geographical area of 1,55,707sq.km., about 41,000km is flood prone in Orissa.

From these data, table 1 and the figure 2, it is inferred that Orissa state is a land of multiple geohazards; however, cyclones cause severe damage than any other hazard, as explained in the following sections.

CYCLONES DEVASTATE ORISSA

Cyclone - An Introduction

Cyclone is a natural phenomenon that occurs due to the variations of pressure and temperature in the atmosphere. A cyclone represents a state of atmosphere, in which there is a low pressure surrounded by a high-pressure zone. Tropical cyclones develop over oceans due to the high sea-surface temperature, which in turn creates instability in the atmospheric pressure. High temperature introduces reduction of atmospheric pressure from normal range (965 milli bar to 1033 milli bar). Decrease of atmospheric pressure below 965 milli bar forms low pressure zone. Being hot and light, the air in the low-pressure zone moves up causing winds from the surrounding high-pressure zone to move inwards of low pressure area. The converging wind is spiral in nature with great velocity and accompanies heavy rainfall. At the center of the cyclone, a zone of calm atmosphere appears which is

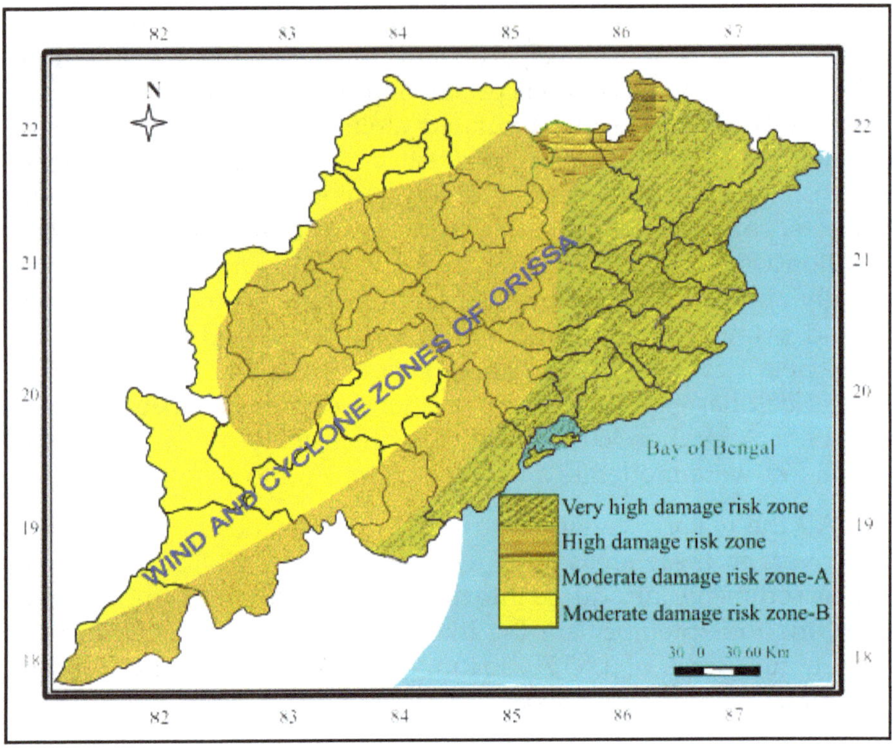

Figure 3. Map of Orissa showing wind and cyclone zones (Coastal track comes under very high damage risk zones.

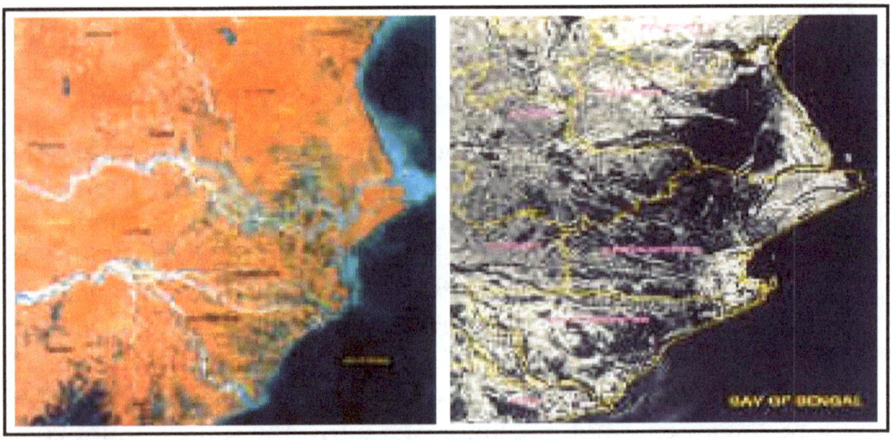

The IRS-1D satellite imagery of coastal Orissa taken on 11th October, 1999 (Before the Super Cyclone)

The IRS-1D satellite imagery of coastal Orissa taken on 2nd November 1999. (After the Super Cyclone)

Figure 4. Imageries of coastal tract of Orissa before and after Super Cyclone

known as "eye". The central calm causes winds to converge on to the center and make the air to rise in the outer boundary of the eye known as "eye wall". A band of weak precipitation, which causes moderate to heavy rain, surrounds the eye wall. In major cyclones, the eye may be 30-50 km in diameter and in average circumstances may take between 30 minutes and 4 hours to pass a point on the surface (Alexander, 1993). Cyclones are classified into four types according to the wind velocity, namely, cyclone, severe cyclone, very severe cyclone and super cyclone.

Cyclones are Regular Visitors to Coastal Orissa

Orissa is the most vulnerable state of India affected by cyclones. Coastal tract of Orissa constituting 9 districts is mainly victim to the cyclonic hazards (Fig. 3) from time to time causing considerable loss of lives and properties. The adjoining districts also suffer from cyclones. The past record tells that during the period from 1891 to 1990, about 1138 cyclone disturbances occurred in the Bay of Bengal region (Table.2). Out of this, Orissa received 273, which is 38.78% of the total number of depressions. Orissa has experienced about 79 storms, the highest among coastal states of India, which is 30.62%of the total number of storms. Orissa has experienced 23 severe storms constituting 13.07% of total number of severe storms.

It was on 29th October 1999, a super cyclone, the worst ever in the country devastated the entire coastal belt of Orissa. The cyclonic storm of magnitude of 250 to 300km/hour whirlwind followed by the 5-8 m high tidal waves and torrential rain struck the lives of people. Although the coastal Orissa is a cyclone prone belt, never in the past did it witness devastation of such a magnitude. Figure 4 shows the status of Orissa coast before and after this super cyclone. The killer cyclone of the century with rare severity caused wide spread havoc in all the coastal districts resulting heavy causalities. It had made the lives of at least 15 million people misearable and death toll had crossed 10,000.

CYCLONE DISASTER MANAGEMENT

Geological hazards are unpredictable severe phenomena of the nature. Disasters cannot be impeded or diverted. Only through effective Disaster Management System that comprises of different components such as proper planning, implementation of remedial measures and forewarning, the potential impacts could be minimized. Disaster management focuses on the various aspects of precautionary measures to mitigate the extremity of the natural calamities upon human being by certain principles, practices and techniques. Disaster management operates on the principles of pre-disaster preparedness, post-disaster settlement and on the overall environmental restructuring of the affected zone. Study of destructive parameters and implementation of suppressant methodos are the tools of effective disaster management.

For effective implementation of disaster management with reference to Orissa coastal zone, lessons learnt from three consecutive disasters in Orissa i.e. the super cyclone in 1999, the severe drought in 2000 and the catastrophic flood in 2001 would be of much help (Beura and Singh, 2006). The disaster management has its scope from early warning system to rehabilitation through rescue and relief operations. All the activities falling under the scope of disaster management are in need of fast and accurate information. Application of information technology in pre and post disaster events can render immense help in managing the distress. Implementation of e-techno-environmental measure has a long-term scope for disaster management. In this context, managing the disasters in general and cyclones in particular can be outlined in to two namely, Pre and Post Disaster Management. These are detailed herein.

Table 2. Historical records of major cyclones along the Bay of Bengal

S.	Date / Year	Place of devastation	Deaths
1	28,29,30 cot.1999	Orissa coast	10000
2	Oct 17,1999	Gopalpur, Orissa	197
3	Nov 5 to 7,1996	Kakinada, Andhra Pradesh	978
4	April 29 - May 02, 1994	Bangladesh	188
5	Dec 01- 04, 1993	Karaikal, Tamil Nadu.	100
6	Nov 11-17, 1992	Tuticorin (Tamil Nadu)	175
7	Nov 11-15, 1991	Tamil Nadu Coast north of Karaikal	185
8	April 29 1991	Chittagang and Kaksbajar, Bangladesh	150000
9	Nov 29 1998	Raimangala, Bangladesh	5708
10	May 04-09, 1990	Machilipatnam, Andhra Pradesh	967
11	Nov 01-09, 1989	Kavali, Andhra Pradesh	69
12	May 23-27, 1989	Northeast of Balasore, Orissa	61

Contd...

Table 2. Contd...

13	Nov 23-30, 1988	Indo-Bangladesh border	2000
14	Oct 31- Nov 3, 1987	Nellore, Andhra Pradesh	50
15	May 24 1985	Chittagang, Kaksbajar and Sandwip Bangladesh	11069
16	Nov 9-14, 1984	Sriharikota and Durgarajupatnam, Andhra	54
17	Oct 16-21, 1982	Sriharikota and dugaraja patinam (A.P)	60
18	Dec 4-11, 1981	West Bengal Coast near Sagar Island	200
19	May 10-13, 1979	Ongole, Andhra Pradesh	700
20	Nov 19-24, 1978	Kilakkarai, Rochemary and Ramnad, Tamil Nadu	915
21	Nov 14-19,1977	Chirala, Andhra Pradesh	10000
22	Nov 8-12, 1977	Tamil Nadu coast south of Nagapattinam	560
23	Nov 15-17, 1976	South Andhra Pradesh coast near Nellorei	30
24	November 3-6, 1976	North of Machlipatnam, Andhra Pradesh	25
25	Sept 6-19, 1976	Southeast Madhya Pradesh	40
26	Dec 1-8,1972	Tamil Nadu coast close to and north of Cudalore	80
27	Nov 15-23, 1972	South Andhra Pradesh coast near Sriharikota	80
28	Sept 7-14. 1972	North Andhra Coast near Baruva	100
29	Sept 7-14, 1971	South Orissa coast and adjoining North Andhra	90
30	Sept 27-Oct 1, 1971	West Bengal coast near Sundarban	160
31	Oct 30, 1971	Paradeep, Orissa	10000
32	May 8 1971	Khulana to Chittagang, Bangladesh	500000
33	May 11, 1965	Nuakholi and Barishal, Bangladesh	19270
34	May 28 1963	Chittagang and Nuakholi, Bangladesh	11520
35	May 30, 1961	Chittagang, Nuakholi and Kaksbajar, Bangladesh	11466
36	Oct 16, 1943	Kandhei, West Bengal	15000
37	May 26, 1941	Nuakholi and Barishal, Bangladesh	75000

Contd...

38	Sept 20-25, 1919	Barishal, Bangladesh	40000
39	Oct 24, 1897	Chittagang and Kutbudia, Bangladesh	175000
40	May 21, 1833	Sagar Island and Chabispragana, West Bengal	50000
41	Oct 11, 1737	Hugli, West Bengal	
42	Nov 10, 1681	Nagpatinam, Tamil Nadu	14000
43	1584	Barakas, Bangladesh	200000

Pre Disaster Management

Unlike other geohazards, cyclones are predictable with fairly accurate timing, location and intensity, as they require considerable time to form and also show visible – either physically or instrumentally signs of their formation. In addition, decades of research had permitted development of sophisticated combination of techniques for early warning system inclusive of oceanographic buoys that document sea-surface and atmospheric temperature, wind speed, radar telemetry and weather satellite that permit enough time lag for observation-modeling-forecasting-tracking-evacuation-relief and rescue measures. Early warning system is the first activity in the pre disaster management. The pre disaster management is equipped with the following systems (after Beura et al., 2002).

- Disaster Prediction System (DPS)
- Disaster Warning System (DWS)
- Public Communication System (PCS)
- Motivation and evacuation system (MES)

Disaster Prediction System : It includes monitoring the data related to severe weather phenomena. DPS should be unambiguous. Otherwise it will dent the trust of common people over the system itself, which in turn, may be disastrous in the event of people turning blind to the warning and the forecast becoming true. The DPS involves the process of receipt of climatic data from various primary and secondary sources and proper compilation, critical study and analysis to get accurate prediction. Parameters responsible for cyclone can be studied precisely under the domain of Disaster Prediction Information System (DPIS – Fig. 5) and then analyzed by software (using fussy logic) to produce the contents of forecasting. DPIS is a subset of Decision Support System in which current and historical data are comprehensively analyzed and explored, creating summaries of the data, in order to support decision-making.

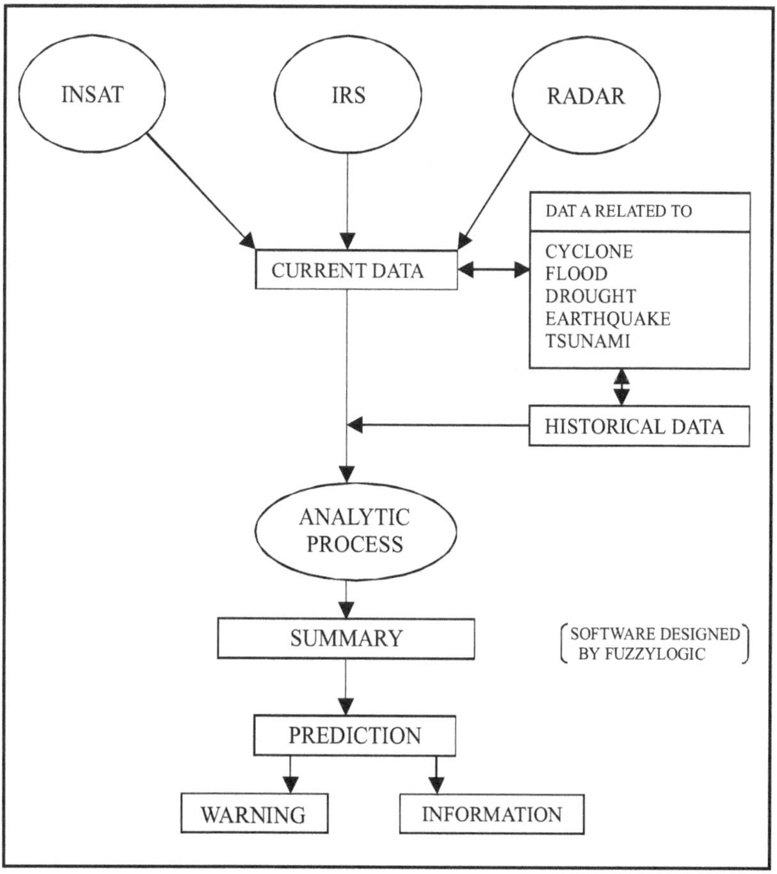

Figure 5. Schematic diagram of DPIS working model

The forecasting data, being a critical tool of DPS, should be spontaneous and error free. Utilization of remote sensing imageries for data inputs into DPS wherein auto-modeling is attempted for accurate and complete understanding of the phenomena. At present a number of data capturing sources are being used for data reception at different centers of the meteorological department to predict the forthcoming disasters. Data from IRS is used for demarcation of disaster risk zones. The meteorological data relayed by INSAT, the world's largest domestic communication satellite system, comprising of INSAT-2B, INSAT-2C, INSAT-2D, INSAT-2E and INSAT-3B are used not only to track the disaster formation but also to issue warning. The WiFS

(Wide Field Sensor) and Short-Wave Infrared (SWIR) band of IRS-P3 have the advantage in giving early warnings of disasters and also assessment of damage.

Irrespective of the satellite data, most important and basic information required for prediction and forecasting are the data on temperature-pressure variations and generation of low-pressure zones in coastal belts and over the seas and oceans. High-powered S-band radars with DVIP facility for estimating data related to different disasters up to 200 km around the radar site are currently in service in India. When cyclone is beyond the range of coastal radar, its intensity and movement are monitored with the help of INSAT and NOAA series of satellites. The INSAT provides every three-hourly cloud pictures over the Indian subcontinent. For precise location, every half-an-hour picture is used.

Data so obtained get accumulated in a database. Data are logically organized into files. Data files are organized in such a way that sorting, indexing and searching are very easy. Data warehouses contain consolidated data from many sources, augmented with summary information and covering a long time period (Tanenbaum, 2000). Warehouses are much larger than other kinds of database; sizes ranging from several gigabytes to terabytes are common. Typical workloads involving *ad hoc,* fairly complex queries and fast response times are important, for which dedicated computer systems with hyper speed processors are available. Maps of physically dynamic areas as rivers, coastal pressure zones and periodic rates and kind of changes of river course can be determined by computerized processing of predictable data so received. Cyclonic process can be monitored and integrated by sequential mapping especially by aerial photographic and other remote sensing data. High-resolution panchromatic and multi-spectral satellite data received from new generation indigenous satellites are getting used as an alternative to aerial photographs.

Disaster Warning System : It is the most important component of disaster management. Warning in advance can accelerate the preparedness program to deal with the havoc. An early cyclone warning system gears up the infrastructural readiness to effectively undertake the evacuation, rescue and relief operation. There are two facets of early warning system. One is the availability of effective techniques to forecast the disaster and the other is effective communication scheme to spread out the message. In the cases of disasters like cyclone, the time available to respond to the hazard is of the order of a few days. Hence, creating awareness on impending hazard, execution of evacuation and being prepared for rescue operations can be made foolproof.

Disaster Warning System acts as a source of disaster information in terms of physical, meteorological and mitigation measures to various nodal centers in addition to common public. India Meteorological Department (IMD) operates cyclone-warning system from the Area Cyclone Warning Centers (ACWC). Cyclone warnings are generated and disseminated through a satellite based communication system called Cyclone Warning Dissemination System for direct dissemination of cyclone warnings to the cyclone prone coastal areas. IMD uses a Limited area Analysis and Forecast System (LAFS), based on an Optimal Interpretation (OI) analysis and a limited area Primitive Equation (PE) model, to provide numerical guidance (Ramakrishnan and Gehrke, 2002). Recently, suggestions are being made to introduce transmission of disaster information to common public and transfer of information from victims through establishment of voice-based software and web-based system in hazard prone regions.

Public Communication System : It facilitates the disaster forecasting and warning to reach the community and individual level (Fig.6). PCS plays a vital role for communicating the disaster and mitigation of devastation during and after the havoc. This system should be inter-communicable and everybody can access real-time information. For that, WAP (Wireless Application Protocol) can be introduced which enables usage of mobile phones to empower the PCS when the print and electronic media get paralyzed. WAP software has been developed for wireless digital communication. WAP technology is based on radio transmission. Because of its popularity, WAP is being frequently upgraded time to time. GPRS (General Packet Radio Service) is an Internet Protocol based packet database system, which enables virtual connections to services, eliminates long dialup delays and allows information such as e-mail without any dependence on major infrastructure. CDMA (Code Division Multiple Access), uses spread spectrum technology, in which many users share the same radio channel simultaneously but are distinguished by pseudo random codes (Rysavy, 2001). It is imperative that geoscientific assessment can be made available to potential users to save countless lives and infrastructure by utilizing these improved technologies in bidirectional information dissemination.

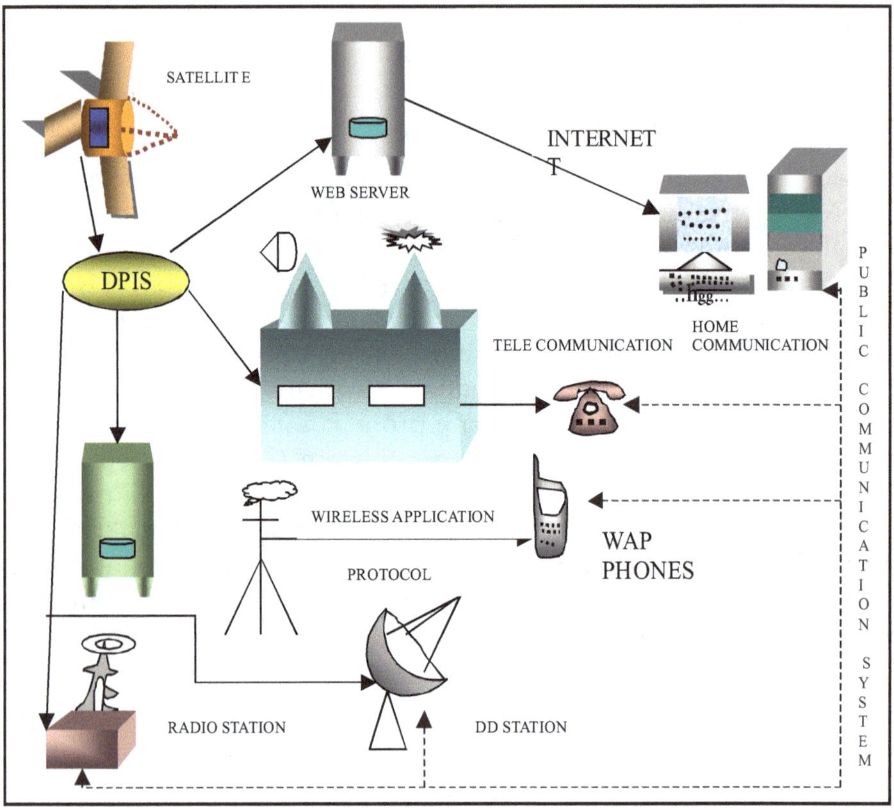

Figure 6. IT based integrated Public Communication System.

Proper and accurate prediction of the disaster leads to the process of evacuation - an important objective of pre disaster management. During this short period of emergency course of action, information technology plays an important role in quick assessment of affected zone. Global Positioning System (GPS) can be utilized successfully to detect such segments. Evacuation is a partial but effective attempt to check the human suffering against nature.

Motivation and Evacuation System : Evacuation is a very important task in the disaster preparedness program. It is well operated with the infrastructural readiness with respect to information system and structural and equipment set-ups. Online GIS assisted with GPS and advanced communication system help the system to locate the segregated patches of risk with shortest root of approach. As evacuating team should avail all sorts of alternate means in terms of vehicle, equipments and shelters etc. to accommodate the displaced ones at the shortest span of time, making available such sophisticated decision support systems would ease the system to take decisions flawlessly.

A prompt warning system may not, in most cases, able to expedite self-evacuation process to safer places. Unprecedented detachment from the place of life and livelihood inhibit common people to leave their houses. They require effective counseling by the team of local volunteers. The volunteers should be trained with the art of risk time motivation, which they can impart it upon people successfully.

Post Disaster Management

Post disaster management includes all efforts that attempt to retrieve affected people and infrastructure from multifaceted calamitous sufferings. Essentially, it targets to restore "pre-event status" as far as possible in terms of livelihood, infrastructure and environment. The efforts include

- Rescue operation
- Relief operation and
- Rehabilitation

Rescue operation : It is considered as an imperative part of disaster management in real time. It requires timely and effective handling of the situation. At times, even a slightest negligence may cost countless lives and precious infrastructure. This tough job demands sophisticated pre-preparedness up to the details and special requirements such as rescue vehicles, power boats, trained manpower, mobile health care unit, portable drinking water tank, permanent and temporary shelter houses, etc.

Rescue operation needs an instant and prompt action package supported by strong infrastructural readiness. Before sending rescue teams, status of the disaster victims at different geographical segment is to be monitored. Geographical Information System (GIS) can provide the devastation picture in terms of location of affected zone, type of losses, amount of losses in terms of property and lives, supply routes, etc. The accumulated data is put into decision support to handle a precise rescue operation.

Relief operation : Relief operation, the challenging part of disaster management can better be materialized by adequate commodity inflow, flexible but genuine distribution. Online transaction process

(OLTP) can play the vital role in managing the relief operations namely deciphering graphic location, selection of mode of transport, distribution status of relief and rescue materials, etc. Victims of natural calamities live on the relief commodities for sustenance till the restoration of normal life. Faulty management in relief distribution is a severe drawback and aggravates the distress of victims. The following management aspects should be taken into account for a better and smooth relief operation.

- Proper co-ordination between the chief relief commissioner and the persons involved in carrying and forwarding the relief.
- Extensive participation of NGOs and voluntary organizations in relief operation.
- Selection, mobilization and pressing in to service of transportation of relief materials to affected areas.
- Regular and continuous distribution of relief materials, dissemination of information on the status of distribution can be of help.
- Surveys at regular interval are to be conducted in the areas of relief operation to know the degree of success and transparency in relief distribution.
- The entire administrative system should be on alert during the relief operation.

Rehabilitation : Rehabilitation is conducted after the incidence of hazards. It is aimed at improving physical, psychological, socio-economical, educational, occupational and environmental conditions of the victims. Proper management in this regard is indispensable for rejuvenating the course of living in true sense. Few important points favoring an effective rehabilitation program are presented herein.

- The cyclone victims should be taken to safe and hygienic place. Sanitation is to be given first priority to avoid endemics.
- Instant and preliminary treatment of the affected people is to be carried out by adequate number of teams consisting of doctors and volunteers.
- For instant and short team settlement, people are to be provided with polythene sheets, blankets, utensils and lanterns or candles etc.

- People are to be provided with monetary help or supplied with construction materials whichever is suitable and essential to construct their houses.
- Psychological disturbances may happen during or after the cyclone. Persons with such conditions should be recognized and admitted for suitable treatment.
- In order to bring back the victims to mainstream they are to be given food for work and alternative way to earn their livelihood has to be provided and or taught.
- Farmers are to be supplied with farming equipment, seeds, fertilizers and pesticides at subsidized rates to restart cultivation.

DISASTER MANAGEMENT SETUP IN ORISSA

It was only after the devastations of super cyclone in Orissa, disaster preparedness setup has been constituted. The Government of Orissa set up an autonomous organization named Orissa State Disaster Mitigation Authority (OSDMA) on 28th December 1999, just after one month of the devastative super cyclone. OSDMA has become the statutory body under the administrative structure of Revenue Department, State Government of Orissa to carry out all types of disaster related assignments. It acts as a nodal agency for multi-tasks like promotion of disaster preparedness programme, coordinating various agencies like UN agencies, international, national and state-level NGO's for aid and involvement, disaster reconstruction work and networking with other similar organizations for disaster management. However, the financial outlay required for long-term measures is vast and the state's resource could hardly accomplish such long-term measures. In this section, the existing command and control mechanism and resources available are detailed.

In Orissa, at Government level, the Revenue Department acts as the nodal agency for disaster management. The field operations are coordinated and supervised by a high level officer known as Special Relief Commissioner (SRC). S/he is given wide powers in terms of administration of relief fund, co-ordination and control of the activities of other departments connected with the disaster management. Divisional commissioners (DC) with similar powers but jurisdictional limits upto each district work under the SRC. DC's prepare contingent plans for their respective districts so as to fit into the state level plan. Sub-Divisional Officers and Block Development Officers work under the control of DC's. This hierarchy goes down to Village Administrative Officer and Field Attendants who are given specific tasks during calamities.

A state level natural calamities committee exists under the chairmanship of the Chief Minister. State level committee includes selected Members of Parliament, Members of the State Legislative Assembly, the Secretaries and Heads of Departments concerned with disaster management. The committee reviews the situation arising out of natural calamities and gives advice on measures to be taken, the nature and quantum of relief to be distributed and the over all policy relating to relief. The Collector is Chairman of district level committees.

The Special Relief Commissioner, before the commencement of the monsoon meets with all Secretaries, Heads of Departments including the Head of Police, Wireless Organization, the representatives of All India Radio, Doordarshan, Posts and Telegraphs Department, Telephone Department, Meteorological Department, Railways and representatives of important voluntary organizations. The meeting also includes the Officer-Commanding and Station Commander, Infantry Battalion stationed at State Capital to participate in the discussions regarding the nature and extent of army support that may be required for rescue and relief operations. Subsequently discussions are held in detail with the authorities of Army and Air Force regarding deployment of army personnel and arrangement for air dropping.

Control rooms are opened to function round the clock from 1st May to 30th November in all districts, at the departments of Government concerned with disaster management. The District Emergency Control Rooms in all 30 districts have been provided with all modern communication equipments to strengthen their workability. The entire district control rooms, block headquarters, few tehsils, Sub-Collectors and few inaccessible Gram Panchayats of 30 districts have been networked with civil VHF communication system. Power backup systems are made ready for all VHF base stations. Some VHF bases are mounted on the vehicles used by disaster managers of the districts. Satellite phones have been supplied to District Collectors of disaster vulnerable districts, Revenue Control Rooms, OSDMA and SRC, Orissa. A test call is made by OSDMA on 1st working day of every month to the Collectors of coastal districts in order to check the alertness of man and machinery. In addition, Free Power Radios are kept ready in Kendrapara, Bhadrak, Balasore, Jagatsinghpur and Puri districts.

Wireless in Local Loop (WLL) Phones are being used by OSDMA and SRC cell at Bhubaneswar.

Relief operation administration is fixed in the hands of Panchayati Raj Institutions in rural areas and in urban local bodies. Government of India and UNDP have taken care to train as many as 195 elected members at district level and 6072 elected members at block levels in Orissa. Disaster preparedness program at village level yields the best preventive measure to protect lives and livelihood. The state government in coordination with United Nations Development Program (UNDP) form teams of volunteers at village level who are being trained to practice and implement village disaster management plan. They carry out the grass root level activities of rescue operation, first aid, shelter management, relief distribution, water and sanitation and rehabilitation functions.

Following the introduction of disaster management in school curricula of CBSE at national level, the Orissa state school education board has also included similar lessons in standard VIII, IX and X. Orissa has also included DM in School curricula. It is made with the intention of sensitizing youngsters about the impending dangers of geohazards and makes them aware about relief and rescue operations.

After the Super Cyclone of 1999, with the bitter experience of mismanagement in rescue and relief operations due to lack of trained manpower, the state government had created Orissa Disaster Rapid Action Force (ODRAF) to assist the civil administration in search and rescue operations, relief line clearance and overall disaster management. The ODRAF units consist of the manpower from Orissa State Armed Police (OSAP) battalions and APR Police. There are about five ODRAF units set up at different corners of Orissa. In addition, Fire Service Units and Civil Defence Organizations are being reinforced to act as smaller specialized response group in search and rescue team.

Orissa Disaster Rapid Action Force (ODRAF) units located at Cuttack, Balasore, Chhatrapur, Jharsuguda and Koraput have been equipped with emergency equipments and heavy-duty machineries namely, hydraulic cranes, trucks, troop carriers, generators, concrete cutting equipments, power saw with portable power units, tree pruner cum chain saw, portable water filter, ramps, telescopic aluminium ladders, life jacket, life buoy, tractors, inflatable tower lighting systems with generators, ambulances, lifting and pumping machines, wire rope and strings, battery chargers, hydraulic tippers, recovery vans and wreckage cranes, submersible pumps, breathing apparatus, diving equipments, gas masks with chemical cadridges and pay loader cum excavator. They are very much essential during the time of evacuation, rescue and relief operation and settlement. Multi purpose cyclone shelters have been constructed in the vulnerable places of coastal Orissa to provide temporary but safe stay during cyclones. With the assistance of World Bank and Chief Minister Relief Fund, about 92 MCS have been completed in six coastal districts within 10 km of high tide level of the Bay of Bengal and handed over to Cyclone Shelter Management and Maintenance Committee. Indian Red Cross Society in collaboration with German KFW and Spanish Red Cross fund has constructed 29 MCS in coastal districts. In cyclone affected districts, the school buildings are newly built or designed and reconstructed in a way to serve as cyclone shelters. About 1303 Primary Schools and 1122 high school buildings are constructed and used as cyclone shelters.

Orissa is one of the 13 states prone to cyclones where Government of India and World Bank sponsored National Cyclone Risk Mitigation Programe is underway and would be completed within the year 2010. The Chief Secretary of Orissa is the Chairman of the Project Steering Committee to monitor the implementation of the program in the state. IMD, Bhubaneswar, communicates warning on impending cyclone and heavy rains. This warning gets conveyed from control Room of Revenue Department to Collector, Collector to Block Department Officer or Tahasildars over phone, fax or wireless. IMD has set up Cyclone Warning Dissemination System (CWDS) Stations at different places to propagate cyclonic warnings. A comprehensive GIS database of the entire coastal district has been created incorporating information such as demography, physical boundaries and geomorphology at the levels of village, panchayat, block and district. The database also includes the information like location of multipurpose cyclone shelters, school-cum-cyclone shelters and VHF base stations. These information would be handy for decision makers for planning and executing relief and rescue operations and saving precious time during emergencies.

CONCLUSIONS

Orissa coast is highly vulnerable to cyclone disasters. The catastrophic characterization of cyclones discussed in preceding sections, points to extreme human suffering due to high density of population and high frequency of occurrence. The operational set up in managing the cyclone disaster has become credible in practice. The task of risk minimization has abridged the participation gap between public and nodal agencies. Socio-environmental reconstruction can be maintained through comprehensive preparedness and participation of people. Agglomeration of mitigation techniques and principles such as DPS, DWS, PCS and MES would certainly be an effective mechanism to develop an integrated disaster management system. In the changing scenario of climates, the management objectives need to be redefined with regard to dynamics of cyclones.

References

Alexander, D., 1993 Natural Disasters, UCL Press Limited, London. 632p.

Building Material and Technology Promotion Council., 1997 Vulnerability Atlas of India (Orissa). Govt. of India. *(Restricted document of internal circulation).*

Beura, D. and Singh, P., 2006 Disaster Management in Orissa coast - An information technology vision. SGAT.Bull. v.7. pp.27-36.

Beura, D. and Singh, P., 2001 Disaster Management with special reference to Orissa Coastal Cyclogenesis. Abst.Vol.Sem.Env.Imp.Ass.Env.Act. Orissa. p.24.

Beura, D., Singh, P., and Acharya, A., 2002 Role of Information Technology in Disaster Management. Proc.Nat.Sem.Role.IT.Geo. Sci. pp.106-108.

Ramakrishnan, R. and Gehrke, J., 2002 Database Management Systems, McGraw-Hill. 936p.

Rysavy, P., 2001 The Road To a Wireless Future. Network Computing, Jasubhai Digital Media, Mumbai. pp.83-90.

Tanenbaum, A. S., 2000 Computer Networks, Prentice-Hall of India. 813p.

❏❏❏

11

An Introduction to Drought

R.SURESH & Mu.RAMKUMAR
Department of Geology, Periyar University, Salem - 636 011.
E-mail: <calcite15@yahoo.com>

Despite widespread occurrence and its potency to inflict heavy causalities and damage to economy, generalizations could not be made to define "drought" – leave alone characterization, development of predictive model and common mitigation measures. Regional variation, be it definition or symptoms or causative factors or consequences and mitigation measures is the hallmark of drought. This chapter is an attempt to present a review on characteristics of drought for the benefit of devising mitigation and reclamation measures.

INTRODUCTION

"Drought" denotes the meteorological, agricultural, environmental, climatic, social, economic and political consequences of a gap between the water availability and demands for the needs of domestic, irrigational, industrial, commercial and recreational consumption in an area over extended period of time. As these demands vary from region to region and depend on too many parameters such as quantum of rainfall, geological and soil characteristics, land use, population, economic activity, climatic conditions, population and culture, drought conditions of an area may not have characteristics of some other region. Such variability makes it difficult to detect. Drought can be defined by rainfall amounts, vegetation conditions, agricultural productivity, soil moisture, levels in reservoirs and stream flow or economic impacts. In the most basic terms, a drought is significant deficit in moisture availability due to lower than normal rainfall (www.cricyt.edu.ar/wdc/paleo/html/drought/drght_alleve.html). Thus, defining drought, mapping drought affected regions and selection of mitigation measures to be adopted have all been tenuous tasks which in turn, together, make management of this geohazard a most dreaded one, often with unsatisfactory results. However, drought is the most serious threat to humanity, environment and economy, as evidenced by the data available in www.preventionweb.net/english/hazards/statistics. This website states that, since the year 1980, there have been 384 drought events world over, which killed 5,58,215 people at an average of 21,470 deaths per year, affected more than 151,80,36,968 people and inflicted economic loss to the tune of 29,28,778 U.S. dollars per year. World disasters report (1994), conclude that, with the exception of civil strife, droughts caused more social disruption than any other natural or non-natural disaster during recent history. World Health Organization estimates that out of total death toll resulted from natural disasters, the drought alone had contributed towards half of it (http://science.jrank.org/ pages/2165/drought-history.html). These observations necessitate thorough understanding on the nature of droughts.

CATEGORIZATION

Droughts differ in many ways from other geohazards. They are slowly accumulating adverse environmental conditions and hence recognition of their onset and dawn is difficult. Their impacts are not restricted to geographic region of occurrence and the effects are cumulative in nature. Many natural processes control and or contribute to dry weather conditions leading to desertification which get aggravated due to many human activities such as land use, culture, demographic patterns, etc. to form severe drought conditions for extended duration. Thus, categorization of drought could be made based on the type of impact as described herein.

As the impacts of drought are initiated from the dryness of a region, drought conditions are defined by the measure of dryness from "average" or "normal" which defines the "meteorological drought". The dryness causes reduction in soil moisture, shortage in rainfall, enhanced difference between actual and potential evapotranspiration etc., which result in crop failure, which is termed as "agricultural drought". Shortages in surface and subsurface water supply for domestic, irrigational, commercial and other activities, owing to extended period of reduced rainfall are termed as "hydrological drought". The combined effects of all these three types of droughts on the well being of socio-economy of a region are termed as "socio-economic drought. In socioeconomic drought, deficiencies of precipitation are linked directly to the rising costs of commodities and services and reduction in supply. The geographic extent of socio-economic drought could not be limited to the region of meteorological, agricultural or hydrological droughts and hence, precise quantification of its impacts is difficult. Jarraud (2006) stated that the interrelationship between these types of droughts has three distinguishing features namely, intensity, duration and spatial coverage. In addition to these four types, the arid regions are considered as the places of "permanent drought".

HISTORICAL OCCURRENCES

Droughts are as old as evolutionary history of human race. First human race that moved out of Africa, about 1,35,000 years ago in search of greener pastures is considered to be due to drought conditions prevalent in places of their dwelling. Droughts are one of the earliest documented hazardous events that find mention in ancient epics of various parts of the World. The webpage http://science.jrank.org/pages/6951/tree.html states that studies of tree rings in the United States have identified 21 droughts including the one as old as 1220 and as recent as 1958. The longest drought identified by this method began in 1276 and lasted for 38 years. The earliest drought recorded in history in the United States was in 1621. From 1876 to 1879, severe droughts in China caused millions of people to die. Russia experienced severe droughts in 1890 and 1921. The 1921 drought in the Volga River basin caused more deaths than the death toll of World War I. During two droughts in 1769 and 1865, an estimated 10 million people died in each of those droughts in India owing to failure of monsoons. One of the worst natural disasters the United States of America experienced was the Dust Bowl drought of 1930's. Slightly less severe drought occurred in U.S.A. during 1950's. However, the one occurred during the years 1987-1989 was the severest in terms of economic loss (http://www.cricyt.edu.ar/wdc/paleo/html/drought/drght_history.html). The droughts in England (1921, 1933-1934 and 1976), Central Australia (1945–1972) and Canada (1983–1985) were also severe. Since 1980 till the end of twentieth centuary, most parts of African continent experienced severe droughts. Since the year 2002, again, many African countries face drought conditions. Since 1994, drought and famine in North Korea inflict heavy human causalities. In 2005, part of Amazon basin had experienced the worst drought in the past hundred years.

All these historic accounts denote that drought conditions and resultant impacts are ubiquitous. Though they frequented every geographic region of the World, the consdquence and damage to lives and economy vary geographically.

IMPACTS

According to www.preventionweb.net, India and China are the two countries that were affected severely during the period of 1980-2005 (four times each) in terms of total number of people affected while the African countries namely Ethiopia, Sudan, Mozambique followed by China lead the pack of total number of death. However, China, followed by Australia and Spain dominate the list of greatest economic damage caused by droughts during the same period. Though the impacts of geohazards are measured in terms of death toll and the quantum of economic damage, the impacts of droughts could not be fathomed in these two terms alone. For example, for the same intensity of drought conditions, the impacts of droughts in developed and underdeveloped countries differ widely. While economic loss results in developed countries, malnutrition and multifold increase in death toll results in underdeveloped countries. Similarly, drought leads to civil unrest, political turmoil, war between adjacent countries owing to mass migration, etc. During drought conditions, priorities shift from public safety, health and social security (http://library.thinkquest.org/16132/html/droughtinfo/effects.html) to measures for survival.

Periods of drought conditions in countries located in tropical region pose additional problem in view of dry soil and wind erosion resulting in destruction of meager forest resources, spreading of infectious diseases among crops as well as forest fauna, flora and human community. Forest fires occur more frequently, affecting atmospheric quality, which in turn, spread health hazards (http://library.thinkquest.org/16132/html/drought.html). The regions dependent on monsoon rains for irrigation face the brunt of extended drought conditions in terms of crop failure, often leading to slowed economic activity, fueling political instability, which in turn become the cause of underdevelopment in many of the Asian and African countries. Reduced productivity (agricultural and industrial), reduction of tourism influx and depletion of resources to maintain recreational activities have all cause reduction of economic activity. Reduction of productivity, either agricultural produce or capital goods, enforces increase in cost of import, again, contributing towards economic damage.

Extended periods of droughts cause desertification, shifting of population to greener pastures and rapid exploitation of natural resources of the newly inhabited regions, leading to disruption of climatic and environmental cycles and processes there (www.naturalhazards.org/ investigate/drought/index.html). These adverse impacts are cumulative in nature and are often irreversible. The reclamation measures are either too costly or not effective making the country or region bankrupt socially, economically and environmentally.

CAUSES OF DROUGHT

As droughts are the resultant of acute water shortage due to lack of adequate rain over extended periods of time affecting environmental, agricultural, industrial, commercial, recreational, economic, social, political and other human endeavors, many parameters including but not limited to, namely, climatic cycles, geographic locations, land use, soil and other physical, geological, environmental systems, vegetational and population patterns, type and intensity of economic activity are all involved, thus making accurate targeting of exact cause of establishment of drought conditions in an area a difficult task. In addition, estimation of the quantum of contribution of all these causes is still more complex. Such intrinsic complexity is the reason why most often drought relief and mitigation measures are destined to be less successful. Following paragraphs present few generalizations.

- Surface and sub-surface water resources are dependent on meteoric precipitation, which is controlled essentially by climatic cycles and geographic location. Though influenced by anthropogenic activities, climate is a natural process which has its own dynamism of shifting principal loci of activity in addition to the duration of cycles (short and long term and their frequency) and relative duration of dry and wet spells which in turn control water available in an area. These cycles are disrupted by the El Niño weather phenomenon, which brings droughts in various parts of the World.

- The water resources are not uniformly distributed all over the World. Though a replenishable resource, the rate of consumption of water far exceeds the rate of recharge, leading to slow development of drought conditions.
- The climatic conditions of continents that are surrounded by oceans are influenced by the sea surface temperature, vertical mixing patterns and temperature of ocean currents. These factors control seasonality, rainfall pattern and formation of dry weather conditions which in turn, if extended for longer duration, lead to desertification and drought.
- In regions where there are few surface water reservoirs, the major source of precipitation is transpiration from plants. As plants become sparse in drought conditions, this source of water for precipitation is diminished. The diminished precipitation further decreases the growth of vegetation, leading to severe drought.
- Despite copious rainfall, the local geography, soil and vegetation conditions and other parameters control the actual availability of water for meeting the demands of natural fauna and flora in addition to anthropogenic activities such as urban centers, irrigational, commercial and industrial purposes. These activities, if not managed properly, bring-in drought conditions.
- Clear-cutting of trees for firewood, overgrazing and over-cultivation, which lead to land degradation, contribute to drought conditions (The Columbia Encyclopedia, 2008). Droughts are characterized by their widespread geographical impact, their longevity, their lack of physical impact other than on vegetation, and their dramatic impact on human population displacement. Droughts are directly associated with the problem of desertification. Desertification is a human-induced problem caused by overgrazing, the cutting of vegetation for fuel, the burning of animal manure and the increased pressure of rising human populations on smaller areas of land leading to non-

sustainable development (www.scienceclarified.com/Di-El/Drought.html). The overgrazing of farm animals, the over-cultivation of farmland and the clear-cutting of forests all contribute to formation of dry, cohisionless soil conditions, adding to the severity of droughts (http://library.thinkquest.org/16132/html/drought.html).

- Elements of a cultural, socio-economic and political nature may contribute to the intensity and duration of the famine, but the ultimate causes are to be found in the environmental problems associated with drought and desertification. The former, through its impact on plants and animals, destroys the food supply, and initiates the famine; the latter, with its associated environmental changes, causes the productive land to become barren, and ensures that the famine will persist, or at least recur with some frequency. Together, drought, famine and desertification have been implicated in some of the major human catastrophes of the past and present and the future (www.encyclopedia.com/doc/1G2-3409400088.html).

INDIAN SCENARIO

India, owing to its unique geography and rainfall pattern, experienced severe droughts during history. The data on historic occurrences of droughts in India, compiled by India Meteorological Department (IMD) state that 44 significant droughts occurred during the past 200 years, at an average of once in four-five years. It is also brought out from this compilation that all the drought years were El-Nino years but all the El-Nino years were not drought years, signifying involvement of many other factors in creating drought conditions in India.

Weather conditions, monsoon patterns and rainfall data of Indian sub-continent were intensively studied by IMD based on which probability of occurrence of drought and its probable frequency were determined. According to this estimate, Assam is the only state wherein possibility of occurrence of drought is in the order of once in 15 years; whereas, the states of West Bengal, Madya Pradesh, Konkan region of Maharashtra, Bihar and Orissa would experience drought once in five years; southern Karnataka, eastern Uttar Pradesh and Vidarba regions could face drought once in four years; Gujarat, eastern Rajasthan and western Uttar Pradesh face the prospect of drought conditions once in three years and the states of Tamilnadu, Jammu and Kashmir, western Rajasthan and Telengana regions would face drought conditions once in two and half years. It is also inferred that most part of Indian sub-continent is prone to drought varying from slight to severe intensity.

MITIGATION MEASURES

Drought is considered to be the most complex and least understood natural hazard, affecting more people than any other natural hazard. Occurrence of drought results in a myriad of economic, social, and environmental impacts in developed as well as developing nations. Vulnerability to drought is determined by many social factors including population, land use, distribution of population, etc. As these factors change over time and thus the assessments on vulnerability would also change. Subsequent droughts in the same region will have different effects, even if they are identical in intensity, duration and spatial characteristics, because societal characteristics will have changed. However, much can be done to lessen societal vulnerability to drought through the development of preparedness plans that emphasize risk management and the adoption of appropriate mitigation actions and programs. Given cognizance to these, this section presents a few mitigation measures.

- By their very nature, droughts impacting geographic regions beyond the core area affected by meteorological drought. For example, the forest fires of Southeast Asia emanated from recent drought has created "Black Smoke" causing health hazards in many countries in this region. Similarly, agricultural drought of major wheat exporting countries cause food shortage elsewhere including affecting UN aid to many of the African countries that depend solely on foodgrain import. Hence, combating drought conditions should be on a global scale.

- As drought conditions are cyclic in nature in tune with climatic rhythms, proactive approach in improving water storage structures, conservation of natural vegetation and encouraging efforts in increasing forest cover, reduction of soil erosion, etc. should be made in order to minimize adverse impacts.

- Strict implementation of land-use controls is required to reduce anthropogenic intervention with natural climatological conditions. Measures like reduction in burining of fossil fuels, discouraging use of land resources for over-grazing and over-cultivation, etc. have to be adhered to.

- The regions of "permanent drought" have to be brought under forest cover through making alternative and salt tolerant crops and also through desalinization projects, etc.
- Similar to coastal environmental regulation acts, land areas also have to be protected based on drought vulnerability classification. Establishment of large database and categorization of land with the help of sophisticated technologies such as GIS (Geographic Information System) based on the critical parameters including demography, type of activities (agricultural, industrial, commercial, recreational, etc.) and other geographic and geological conditions has to be undertaken. Master plans at national, state, district, taluk, block and village levels has to be created with this database and all the future developmental activities have to be undertaken in resonance with the suggestions of scientific rationale only. Legal sanction is also required for strict implementation of this master plan.
- Above all, creating better awareness on causative factors, mitigation measures and community participation in combating measures is necessary.

References

http://library.thinkquest.org/16132/html/drought.html.

http://library.thinkquest.org/16132/html/droughtinfo/effects.html.

http://science.jrank.org/pages/2165/drought-history.html.

http://science.jrank.org/ pages/6951/tree.html.

Jarraud, M., 2006 Drought monitoring and early warning: Concepts, progress and future challenges. Concept paper of World Meteorological Organization. 26p.

Somerville, C. M., 1986 Drought and aid in the Sahel. Westview, Boulder.

The Columbia Encyclopedia, 2008.

World Disasters Report, 1994 International Federation of Red Cross and Red Crescent Societies, Geneva, Switzerland.

www.cricyt.edu.ar/wdc/paleo/html/drought/drght_history.html.

www.encyclopedia.com/doc/1G2-3409400088.html.

www.naturalhazards.org/investigate/drought/index.html.

www.preventionweb.net/english/hazards/statistics/.

www.scienceclarified.com/Di-El/Drought.html.

❑❑❑

12

GIS Based Model for Drought Assessment

***P.H.SHIVA PRAKASH, **P.K.GARG & **S.K.GHOSH**
*Department of Civil Engineering, Siddaganga, Tumkur - 572103.
**Geomatics Engineering Research Group, Department of Civil Engineering, IIT - Roorkee, Roorkee - 247 667. E-mail: <gargpfce@iitr.ernet.in>

This chapter explains how natural disasters are major catastrophic events, which are often aggravated by human intervention, resulting in adverse conditions affecting both natural resources and human habitats. Though drought is as disastrous as any other geohazard, it differs from other geohazards such that it is a 'creeping phenomenon', making its onset and end difficult to determine. Only a continuous monitoring of critical parameters ranging from crop failure to joblessness and shifting population, etc, could indicate drought in an area. In this chapter, we demonstrate a GIS based drought forecasting and monitoring model taking Gubbi Taluk of Karnataka as a case study. The approach includes creation of a spatial database of 17 critical parameters and their integration in GIS environment by developing a suitable rating and ranking scheme for the generation of drought severity map. The results have shown that drought could be monitored with the studied parameters and if suitable remedial measures are implemented, their effectiveness could also be monitored with this GIS model.

INTRODUCTION

India is a large country in terms of geographical area, exhibiting greater agro-climatic variation. The country is susceptible to several natural disasters, which are a major constraint to developmental activities. The statistics of the disaster events show an alarming trend (Manikiam, 2003). Disasters also cause untold misery and havoc to lives and livestock (Rao, 2000). It is estimated that 4 billion people – one half of the world's population will live under conditions of severe water stress by the year 2025, with conditions particularly severe in Africa, Middle East and South Asia including India (Diwan, 2002). Drought is a phenomenon that "creeps in" due to scarcity or non-availability of water for agricultural and domestic consumption in an area. Drought can occur under different climatic conditions but its recurrence and effects are more pronounced in arid and semi-arid situations (Singh, 2003). Before the drought, the ground water level sinks and natural water sources including rivers dry-up (Prakash *et al.*, 2003).

The problems of drought-prone regions in India vary in magnitude, temporally and spatially. One of the worst natural calamities that affect India is the large-scale incidence of drought during the southwest monsoon season (June to September). Semi-arid regions of India, mostly rainfed, face the frequent drought hazard resulting in low and uncertain crop yields. For most of the dry crops, the yields are invariably dependant on residual soil moisture storage (Dhopte, 2002), resulting in crop failure and finally famine. Even though India has a long history of drought events in the past, it lacks proper drought management strategy at national level yet. In the absence of drought severity map, no drought relief measure and management would be meaningful.

Satellite data have been used frequently in the past for the study of disaster and land degradation. The obvious advantage of this data product is its large areal coverage, availability in digital form and multi-temporal and multi-spectral analysis for faster and accurate classification. For qualitative assessment of drought, it is necessary to integrate and study the combined effects of terrain, meteorological and land characteristics of the area, for which GIS is indispensable. GIS in conjunction with remote sensing and ancillary data, can be used to identify drought prone areas. Once the drought-affected areas are identified, their representation can be stored conveniently in GIS databases. GIS technology provides a powerful tool for displaying outputs and permits user to visualize the geographic distribution of impacts and allows the user to perform a quick graphical sensitivity analysis of the factors affecting drought (Gupta, 2003). Thus, this study attempts to demonstrate the ability of GIS in recognizing drought prone areas with the help of remotely sensed data and conventional data, taking Gubbi Taluk of Tumkur District, Karnataka State as a study area.

STUDY AREA

The study is carried out for Gubbi Taluk of Tumkur district, covering an area of about 1223 km^2 (11.54% of Tumkur district). It consists of seven hoblis (a hobli is a cluster of villages) comprising 348 villages (Zilla Parishad, 2001). It has a semi-arid climate, undulating terrain, dry land cultivation (mostly rainfed) and barren (scrub) land and fallow and degraded grazing lands. The average annual rainfall is 572.4 mm based on 1901-1970 rainfall data (Census Report, 1971). The major land use/land cover includes mixed cropland dependent on rainfall, water tanks, vegetation (uncropped and plantation), barren land consisting of poor to nil vegetative cover and rocky surface. The region shows dry irrigation wells due to over exploitation of ground water.

GENERATION OF THEMATIC LAYERS

The data used for this study comprise remotely sensed data and conventional data from different sources. Remotely sensed data consists of six IRS 1C LISS-III images of October 1996, September 1999 and October 2001 (three monsoons) and March 1997, March 1999 and March 2001 (three non-monsoons). Conventional data include rainfall data for fifty years (1951 to 2001), data of ground water level fluctuation during the years 1970 to 1998, soil properties for the period from 1994 to 2000, topographical maps (57C/10, 57C/11, 57C/14, 57C/15, 57C/16, 57G/3 and 57G/4), climatic data namely temperature, evaporation, relative humidity, evapo-transpiration and wind speed, data from census report and crop statistics (Census Report, 1971, 1981, 1991 and 2001).

Seventeen thematic layers were generated and used to create a database in GIS for drought analysis (Shiva Prakash, 2006). From topographic maps, *Taluk* limit, contours, drainages, water bodies and forest/plantation layers were generated. The village limits were incorporated from revenue map. Rainfall, temperature, relative humidity, evaporation and evapo-transpiration, ground water, soil, land capability and land irrigability layers were generated from climatic and published data. Using census data, total irrigated land, tank irrigated land, cultivators, agricultural labor, population density and village amenity layers were generated. Thematic layers on geomorphology and hydrology (water bodies, streams etc.) were also generated using topographic maps and satellite images.

ANALYSIS OF REMOTE SENSING IMAGERIES

From remotely sensed data, after georeferencing, a series of new data were generated viz., Normalized Difference Vegetation Index, (NDVI), Principal Component (PC) and Tasseled Cap Transformation (TCT). All the 12 bands were used to obtain land use classes by adopting supervised classification using ERDAS Imagine software. Since the study area has a mixed cropping pattern, only cropland was identified. Water and moistened water body were also very distinctly identified. Barren land was easily identified. The rocky outcrops and vegetation cover were clubbed to form single category. The water tank was the only identifiable water category since no river exists in the study area. Major part of the study area is occupied by *kharif* crop, fallow land with or without scrubs and degraded grazing and barren land. Ground truth has been collected at specific locations during field visits made in the years 2002 and 2003. The accuracy of classification of the images is found to vary between 89 to 97% (Shiva Prakash, 2006). The results of land use classification for monsoon periods are given in Table 1 which shows that the study area had experienced severe drought during the year 2001 as indicated by least count of extent of water body and higher extent of barren land.

Table 1. Landuse statistics during monsoons of 1996, 1999 and 2001

Class	1996 Oct.	1999 Sept.	2001 Oct.
Barren area (%)	23.66	45.74	28.82
Cropped area (%)	17.32	18.47	20.47
Water body (%)	1.42	1.23	0.95

DEVELOPMENT OF A GIS MODEL

Through models, it is possible to process various scenarios and to predict potential influences and developments. The modeling procedure therefore requires an understanding on the processes triggering a system so that these processes may be described or simplified adequately. All the layers, as described above and given in Figure 1, are digitized, edited, topology built and exported in Arc View. By using Spatial Analyst module of Arc View, integration of multi-thematic information, overlaying, analysis, correlation and generation of drought severity map (Fig. 1) were carried out (Shiva Prakash, 2006).

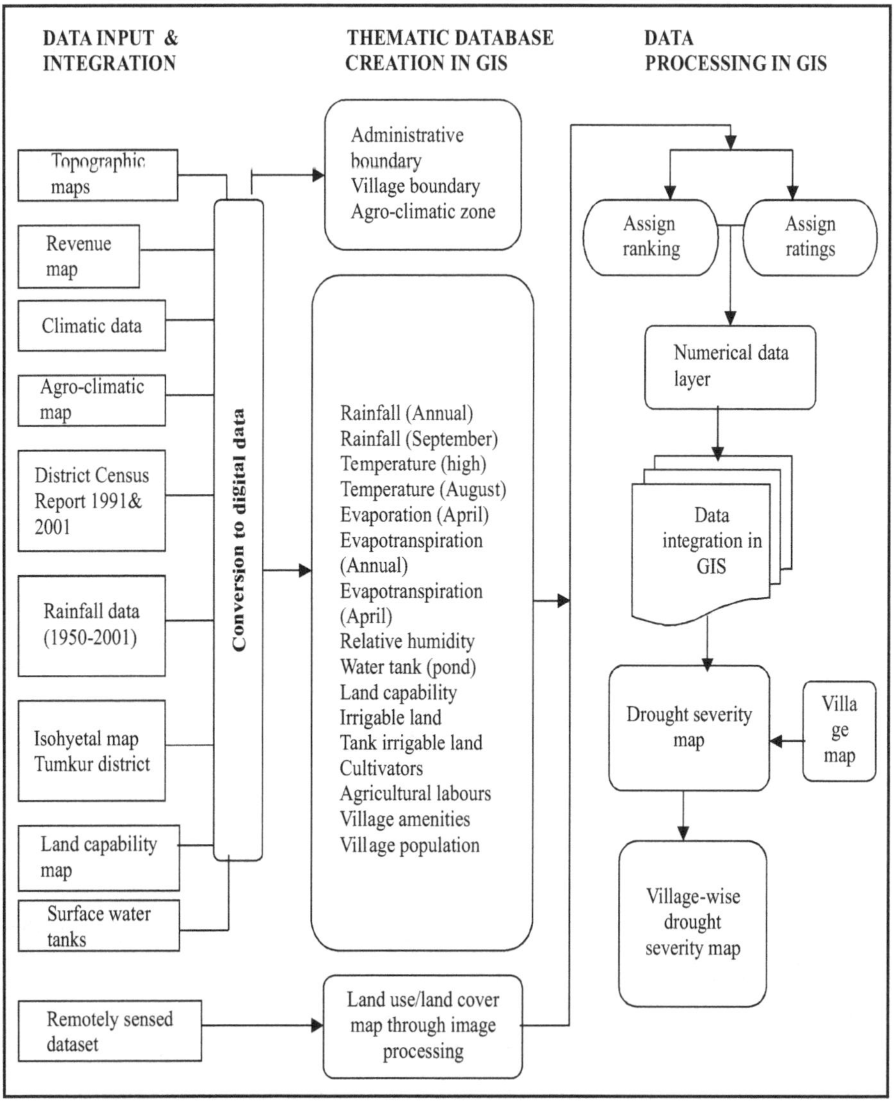

Figure 1. Methodology used in GIS (*after* Shiva Prakash, 2006)

Ranking and Rating of Studied Parameters

Assigning ranking to parameters of drought by arbitrary means is likely to induce subjectivity in their rankings, which must be removed in order to eliminate bias in assigning the rankings (Nagarajan, 2003). Taking cognizance of this opinion, ranking was assigned to each parameter based on its importance in the event of occurrence, together with the rating for the individual classes that denotes the event intensity. Ranking and rating is a technique, which provides a mathematical method to analyze and construct a map from overlays or data from other related maps. Ranking of each parameter was computed using Saaty's pairwise comparison matrix. Table 2 presents the details of various parameters and weights that indicate the drought intensity. The weights assigned to reflect importance of each parameter are shown together with the rating for individual classes (5 being the highest and 1 being the lowest drought intensity).

Application of Model

The arithmetic overlay approach built into Arc View Model Builder has been adopted for integration of input data layers. The rating of each parameter was multiplied by its rank and the sum of the cumulative values of all parameters was utilized for categorization of drought into different classes. The drought severity map for October 1996 was generated by running the model with the landuse/land cover layer of October 1996 and August month temperature layer and all other layers as shown in Table 2. The drought severity map generated by this model is presented in Figure 2 that shows only three drought classes, viz., mild, moderate and severe.

The procedure described above was followed for the generation of remaining season-wise drought severity maps by considering the corresponding landuse/land cover layers, temperature and evapo-transpiration. The villages were identified according to the drought classes by overlaying the village map on the drought severity map. Like-wise, running the corresponding drought model generated the drought severity maps of the remaining seasons and thus the village-wise drought identification was made. Table 3 presents the abstract wherein the drought classes are compared.

Table 2. Ranking and rating for various parameters

Sl.	Parameter	Ranking	Class	Rating
1	Annual Rainfall (mm)	0.165	650.0-658.0	5
			658.0-666.0	4
			666.0-674.0	3
			674.0-682.0	2
			682.0-690.0	1
2	Monthly Rainfall (mm) (September)	0.1517	160.0-164.8	5
			164.8-169.6	4
			169.6-174.4	3
			174.4-179.2	2
			179.2-184.0	1
3	Max. Temperature (°C) (April)	0	34.20-34.29	1
			34.29-34.39	2
			34.39-34.48	3
			34.48-34.58	4
			34.58-34.68	5
4	Monthly Temperature (°C) (August)	0.1062	27.40-27.46	1
			27.46-27.53	2
			27.53-27.59	3
			27.59-27.65	4
			27.65-27.72	5
5	Evaporation (mm)	0.0769	8.60-8.68	1
			8.68-8.76	2
			8.76-8.84	3
			8.84-8.92	4
			8.92-9.00	5
6	Max. Evapotranspiration (mm)	0.0526	164.00-164.64	1
			164.64-165.28	2
			165.28-165.92	3
			165.92-166.56	4
			166.56-167.20	5
7	Monthly Evapotranspiration (April) mm	0.0522	159.00-159.96	1
			159.96-160.92	2
			160.92-161.88	3
			161.88-162.84	4
			162.84-163.80	5
8	Relative Humidity (%)	0.0394	36.80-37.28	5
			37.28-37.76	4
			37.76-38.24	3
			38.24-38.72	2
			38.72-39.20	1
9	Water tanks	0.0451	0.05-17.50	5
			17.5-35.0	4
			35.0-52.5	3
			52.5-70.0	2
			>70.0	1

Contd...

Table 2. Contd...

10	Landuse/land cover map (1996 October)	0.0342	1	1
			2	1
			3	2
			4	2
			5	2
11	Land Capability	0.0301	1	1
			2	2
			3	3
			4	4
			5	5
12	Irrigable Land (ha)	0.0267	0-50	5
			50-100	4
			100-150	3
			150-200	2
			>200	1
13	Tank Irrigable Land (ha)	0.0255	0-25	5
			25-50	4
			50-75	3
			75-100	2
			>100	1
14	Cultivators	0.0253	0-50	5
			50-100	4
			100-150	3
			150-200	2
			>200	1
15	Agricultural Labours	0.0152	0 -50	5
			50-100	4
			100-150	3
			150-200	2
			>200	1
16	Village Amenities	0.014	1	1
			2	2
			3	3
			4	4
			5	5
17	Population	0.0136	0-500	1
			500-1000	2
			1000-1500	3

Table 3. Villages affected by drought during 1996-2001

S. No	Drought class	October 1996 No. of villages (Study area %)	March 1997 No. of villages (Study area %)	March 1999 No. of villages (Study area %)	September 1999 No. of villages (Study area %)	March 2001 No. of villages (Study area %)	October 2001 No. of villages (Study area %)
1	Mild	52 (14.17)	14 (4.25)	0 (0.00)	35 (14.55)	0 (0.00)	34 (12.62)
2	Moderate	238 (67.96)	317 (90.90)	262 (76.93)	307 (83.76)	270 (78.14)	306 (85.31)
3	Severe	58 (17.87)	17 (4.85)	86 (23.09)	6 (1.69)	78 (21.86)	8 (2.07)

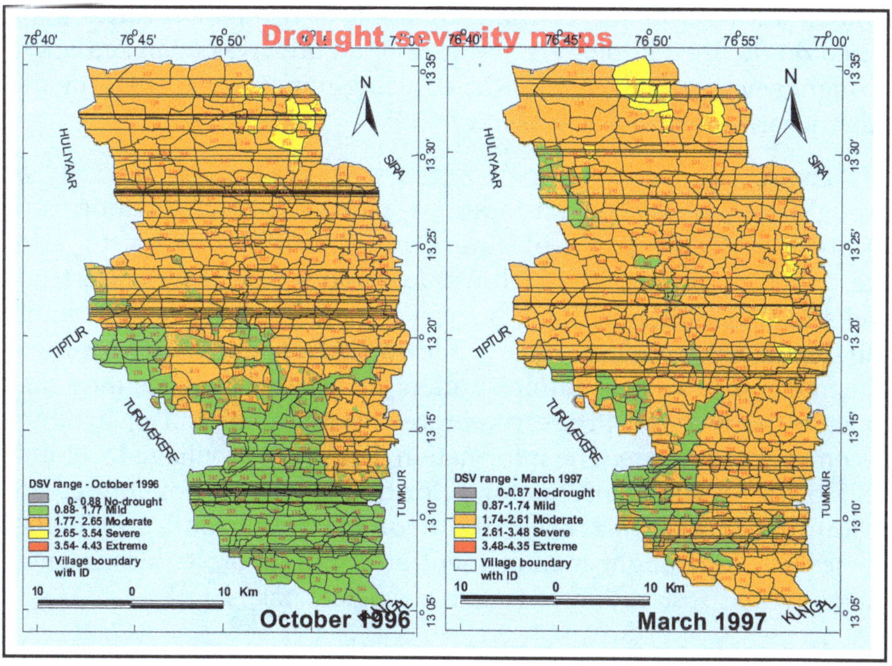

Figure 2. Drought severity maps of October 1996 and March 1997

It is found that no-drought and extreme drought conditions are totally absent during the study period. An average of about 14% of mild drought is observed during October and September in 1996, 1999 and 2001. The number of villages affected during October 1996 is 52, which are the highest. It is observed that that moderate drought is more predominant (covers from 68 to 91% of study area) compared to mild and severe droughts. The number of villages affected during October 1996 is 238 while during September 1999 the number of villages has increased to 307. It is found that even in October 2001 moderate drought accounts for 85.31% affecting about 306 villages. The extent of severe drought varies from 1.7 to 23% affecting villages from as low as 6 to as high as 86. It is observed that the drought condition varies in the spatial extent as well as the number of villages that are affected by drought.

CONCLUSIONS

A methodology for drought analysis has been developed based on seventeen parameters. The approach included analyzing the remotely sensed data to derive land use information, creation of spatial database and its integration in GIS environment by developing a suitable rating scheme for the generation of drought severity map. Based on the methodology, a test case of an area was analyzed and a drought severity map was produced using remotely sensed data and other information with the help of GIS.

The drought severity map would help to point out severity of drought experienced at the locations/places. This information is of utmost importance to planners and administrators to take precautionary measures, which would avoid a much more disastrous and serious situation at a later stage, if neglected. The local inhabitants can use the results of this study to alleviate the continuation of drought phenomenon. The local administration can take the required measures to save the life and property before the phenomenon of drought becomes more severe. The information generated at village-level unit (micro level) can be useful for prioritizing villages for immediate drought combat measures like drought relief, soil and water conservation, drought management and mitigation, environmental planning and restoration of fragile ecological balance.

This study demonstrates the application of an integrated remote sensing and GIS based methodology for drought severity mapping and its management. Since the satellite images are available at regular short time intervals, these can be used for the prediction of both rapid and slow events of droughts. These images can assist in damage assessment and aftermath monitoring, providing a quantitative basis for relief measures. The model generated may guide for a quicker approach to identify prevailing drought-hit villages along with the extent of barren land. The study on drought spread and intensification is of great practical relevance to agricultural importance in general, and especially for planning drought management and combating drought.

References

Census Report., 1971, 1981, 1991, 2001 Government of India, Bangalore.

Dhopte, M. Arvind., 2002 Agrotechnology for dry land farming. Scientific Publishers, Jodhpur. pp.1-10.

Diwan, P.L., 2002 Water environment and drought. Proc. Nat.Sem. "Water and Environment-Issues and Challenges" October 2002, IIT-Roorkee, India. pp.21-185.

Gupta, A., 2003 A tool for disaster managers. Geospatial Today. pp. 37-40.

Manikiam, B., 2003 Remote sensing applications in disaster management. Mausam. v.54. pp.73-78.

Nagarajan, R., 2003 Drought Assessment, monitoring, management and resources conservation. Capital Publishing Co., Bangalore. pp.1-32.

Prakash, V.S., Srinivas Reddy, G.S. and Prabulingappa, A.V., 2003 Drought monitoring system in Karnataka state: Present practice and future plans. Proc.Nat.Workshop on Drought Mitigation, May 2003, Bangalore. pp.1-18.

Rao, D.P., 2000 Geographic data needed in the interpretation of Indian satellite based remote sensing data: Opportunities and realities. Curr.Sci. v.79. pp.468-473.

Shiva Prakash, P.H., 2006 Drought assessment using remote sensing and GIS techniques. Ph.D. dissertation submitted to Indian Institute of Technology-Roorkee, (unpublished).

Singh, S., 2003 Drought and its management in India. Short Term Course on Drought Analysis and Management, July 2003, IIT-Roorkee. pp.1-15.

Thiruvengadachari, S. and Gopalakrishna, H.R., 1993 An integrated PC environment for assessment of drought. Inter.Jour.Remote Sens. v.14. pp.3201-3208.

Zilla Parishad., 2001 Report of Village Information of Zilla Parishad Tumkur. Govt.of Karnataka. pp.4-40.

13

Flooding - A Manageable Geohazard

Mu. RAMKUMAR
Department of Geology, Periyar University, Salem - 636 011.
E-mail: <muramkumar@yahoo.co.in>

This chapter portrays flooding as a geohazard that can be managed with ease through simple and effective measures. Flooding is a recurrent geohazard that affects vast lands in India and elsewhere. Unlike many other geohazards, flooding event and probable area to be affected could be predicted with fairly accurate levels and hence, proper safety measures could be taken easily. However, as flash floods could not be predicted and the nature of flooding hazard that it brings in added danger of health hazard, crop damage and cutting off of supply and communication networks, flooding is as serious threat as any other geohazard. Scientifically designed urbanization and flood control measures are suggested for minimizing the potential impacts of flooding.

INTRODUCTION

Flooding is defined as a temporary submergence of a large area (that normally remains dry) under water by overflowing rivers and reservoirs as a result of heavy rains, high winds, cyclones, storm surges, tsunami, melting of snow or dam burst (Arya *et al.* 2005). The term "temporary submergence" denotes a time period ranging from few hours to about few months depending on type and location of flooding. Although flooding is a natural phenomenon that plays a major role in replenishing wetlands, recharging groundwater and supports agriculture, extreme demands on natural resources due to population growth, people and their property move closer to water bodies, which in turn increases the risk of life and property loss due to flooding (World Meteorological Organization, 2005). Unlike many other geohazards, flooding could be forecast except in the case of flash flood (which is defined as submergence of vast lands under water due to heavy rainfall or cloud burst within six hours of downpour) and the probable area to be affected could be predicated. However, flooding is a serious threat to lives and property if not managed properly as they damage crops and livestock and affect infrastructure, in addition to inflicting health hazards. Sinha and Jain (1998) describe flooding as probably the most recurring, widespread, disastrous and frequent natural hazard of the World. India is one of the worst flood-affected countries, accounting one fifth of global death count due to floods.

CAUSES OF FLOODING

In a natural hydrological system, rainwater either percolates downward to get stored as groundwater and or moves over surface as surface runoff to get collected by rivers. When the rainfall exceeds the quantity that can be collected into groundwater storage aquifers and carrying capacity of rivers, flooding occurs. Thus, excessive rainfall beyond the capacity of river channels and reduction of volume of water percolated to underground are the main causes of flooding. Simple, as it may seem to be, the real cause of flooding is that, with the growing population and extensive usage of natural resources such as land by the human community, reduces the infiltration while increasing the surface run off and rate of run off through urbanization, deforestation and reduction of volume of river channel, increasing the frequency and magnitude of flooding.

Other reasons of flooding include melting of snow, cyclones, storm surges, cloudbursts, dam or reservoir failures and blockage of natural drainage systems due to siltation, wastes and landslides. Marine waters flood the coastal low lands during cyclonic storms in addition to surface and stream flow of meteoric water. The website www.dartmouth.edu/

~floods/archieveatlas, states that more than 1500 flood events all over the world per year result by heavy rain, brief torrential rain and rains associated with tropical cyclones while the other causes such as monsoonal rain, snowmelt, dam or levee breach, ice jam/breakup, extra-tropical cyclone, tidal surge and avalanche constitute the rest of about 300 flood events.

There is another significant characteristic of river systems that they shift their courses depending on quantum of flow, slope, lithological and tectonic conditions, etc. The courses of modern rivers have all been regulated through settlements along and dams across. When the river channel overflows through high run-off, it tends to recapture its original course in which it was running decades if not hundreds of years back, causing extensive flooding of areas previously unheared of flood disaster. The recent flood tragedy in Bihar is a typical case of this phenomenon of a river system. The Kosi River has flooded the areas, which have never seen a flood for the past 150 years. Post flood analysis has shown that the river has recaptured its paleocourse after a span of more than hundred years, which in turn became a site of human habitations during this period.

Post-flood event analysis of Australia, Europe, Canada, Mexico, Cambodia, Burma, Vietnam, Thailand and Canada had revealed that among all these causes, human interference over natural flow systems by deforestation in catchments and urbanization, industrialization and developmental activities in flood plains wherein river channels were distributing floodwater and sediment is the most serious cause. In a classic book, Parker (2000) has stated that the rapid growth of urban areas leads to unplanned haphazard patterns of development with significant pressure put on existing open spaces of flood plains. According to him, it is not merely the existence of development at risk in flood plain that creates hazardous situation. The process of urbanization itself increases flood magnitude and frequency by altering local hydrological characteristics. First, urbanization inhibits infiltration by creating impermeable surfaces such as roads and rooftops. As a direct result, a higher percentage of water enters streams as overland flow. Second, this increased runoff reaches rivers faster because of smoother surfaces and less friction. Thus, lag time between rainfall and discharge into river is decreased while the quantum of discharge is increased. Development in flood plains along and across can constrict or restrict the flow of water in the channel. Bridges reduce the natural carrying capacity and provide barriers upon which debris can accumulate. In addition, development along the rivers can intrude on the river and restrict flow. Thus, a combination of natural and anthropogenic causes contributes towards development of a natural process - "flooding" into a "disaster" or "geohazard".

EFFECTS OF FLOODING

Flooding is a natural process that effects recharge of underground reservoirs of flood plains, distribution of sediments (Photograph 1), fertile soil and nutrients to natural ecosystems. Ancient civilizations and modern cities flourished along river channels and adjacently located plain lands only because of availability of cultivable fertile lands, potable water and navigable river channels for transport of people and goods. However, with the population growth, natural drainage systems are destructed. Urbanization and deforestation of catchments result in accelerated erosion (Photograph 2) of headlands, siltation of river channels and reduction of carrying capacity of natural river systems, which in turn cause flooding.

Flooding results in many adverse impacts namely, damage to lives and property, clogging drainage systems of cities, contamination of water bodies and supply lines, destruction of croplands and disruption of transport (Photograph 3), communication systems, distribution of toxic materials into natural ecosystems and human inhabitations. Prolonged submergence of inhabited regions (Photograph 4), forests and cultivated lands brings in communicable diseases and health hazards, creating epidemics among vulnerable flora and fauna including humans. The impacts of flooding are classified into three major categories; namely, primary, secondary and tertiary. The primary impacts are those directly caused by the flood itself, such as submergence, deposition of large quantity of sediment (Photograph 1), erosion (Photograph 2), and physical damage by moving water (Photograph 3), etc. The secondary effect is the result of primary impact, such as ruining of cultivated lands, contamination of water bodies, disruption of communication network, waterborne diseases etc. The change of river courses and permanent damage to infrastructure are tertiary impacts.

Photograph 1. Thick pile of sands deposited over flood plains and mangrove swamps due to heavy flooding by the Krishna River. Such natural process of sediment deposition in floodplains constructs land and form natural levee along riverbanks. However, the same process of sediment deposition in flood plain, if it is inhabited, is a hazard that causes damage to infrastructure and urban centers.

Photograph 2. Vainateyam branch of River Godavari, eroding severely its banks. Note the presence of vertical cliffs along riverbank and the occurrence of coconut trees within river channel. It indicates that the former bank regions were submerged under the river channel.

Photograph 3. Disruption of transport routes by flooding in a distributary of Cauvery river

Photograph 4. Flooding in a distributary of Cauvery river. Note the occurrence of raised bank along inner curve of the meander (left side of the photograph) and the occurrence of landscape at the water level (right side of the photograph). Outer curves of meanders are always erosional and hence landuse planning and flood control measures should take into account these characteristics of river systems.

The direct and indirect costs of reclamation efforts on flood affected region always go to astronomical levels and thus, flooding is bane to economy too. It is also opined by economists that the highest economic loss due to natural disasters shifts from hurricanes to wild fires to floods in recent times, signifying the destructive nature of flooding hazard than any other geohazard. The following table (Table 1) gives an impression on the magnitude of damage inflicted by floods alone, all over the world during the period from the year 1980 to 2000. These statistics are based on the official reported figures and the actual figures may be still higher. Among countries, China ranks first in most number of floods, most number of people affected and maximum economic damages during the reporting period.

Table 1. Statistics of flood damages all over the world during 1980-2000

Particulars	Statistics
Number of flooding events	2,533
Number of people killed	183,524
Average people killed per year	7,059
Number of affected people	2,591,953,380
Average number of people affected per year	99,690,515
Total economic damage (US dollars)	351,735,158,000
Economic Damage per year (US dollars)	13,528,275,000

Source: www.preventionweb.net/english/hazards/statistics/

LOCALES OF FLOODING AND THE INDIAN SCENARIO

A report of United Nations (1976) states that flood plains, lower river terraces and coastal plains are most vulnerable to flooding. Flooding occurs in lowlands. "Lowland" is a relative term with reference to the difference of relief between water level of water body and the adjacent land. In this context, all the lands adjacent to river channels, river deltas, land region surrounding water bodies such as dams, lakes, etc, are susceptible to flooding. The coastal lands located under the influence of tide, cyclonic storm surge etc., also get flooded during extreme weather. Based on these surmises, it is estimated that about 40 million hectares of lands (12 % land area) of India are prone to flooding (www.cwc.nic.in). In addition to the causes listed in preceding section, flooding in India is due to a specific reason that, although India receives an annual rainfall of 1200 mm, 85% of which is concentrated in 3-4 months. Such a peculiar phenomenon creates a situation that these copious flows could not be transported within existing river channels and underground aquifers. That is, the carrying capacity of the river channels rate of percolation to underground are dismal when compared with the rate of inflow, causing flooding a seasonal and recurrent problem in India. Added to this problem, cyclonic storms that normally coincide monsoonal seasons aggravate the flooding situation. Thus, flooding occurs in lands adjacent to river channels and the downstream regions located adjacent to coastal regions are doubly affected by storm surges and marine inundation.

The website of Geological Survey of India (www.portal.gsi.gov.in/portal/) lists that flood plains and delta of Ravi, Sutlej, Yamuna-Sahibi, Ganga, Gandak, Ghaggar, Teesta, Kosi, Brahmaputra, Mahananda, Mahanadi, Damodar, Mayurakshi, Godavari and Sabarmati and their tributaries and distributaries are prone to flooding. Most of the flood affected areas lie in the Ganga basin, Brahmaputra basin (comprising of Barak, Tista, Torsa, Subansiri, Sankosh, Dihang and Luhit), the northwestern river basin (comprising Jhelum, Chenab, Ravi, Sutlej, Beas and the Ghagra), peninsular river basins (Tapti, Narmada, Mahanadi, Baitarani, Godavari, Krishna, Pennar and the Kaveri) and the coastal regions of Andhra Pradesh, Tamilnadu, Orissa and Kerela. Assam, Uttar Pradesh, Bihar and Orissa are prone to floods. From the Table 2, it is evident that an average of 7.181 million hectares of land get flooded annually in India and also that the year 1978 had recorded greatest ravage, followed by the year 1988.

Table 2. Data on flood affected lands in India during 1953-2000

Year	Land affected (Million hectares)	Year	Land affected (Million hectares)	Year	Land affected (Million hectares)
1953	2.29	1969	6.20	1985	8.38
1954	7.49	1970	8.46	1986	8.81
1955	9.44	1971	13.25	1987	8.89
1956	9.24	1972	4.10	1988	16.29
1957	4.86	1973	11.79	1989	8.06
1958	6.26	1974	6.70	1990	9.30
1959	5.77	1975	6.17	1991	6.35
1960	7.53	1976	11.91	1992	2.65
1961	6.56	1977	11.46	1993	11.44
1962	6.12	1978	17.50	1994	4.81
1963	3.49	1979	3.99	1995	5.25
1964	4.90	1980	11.46	1996	8.05
1965	1.46	1981	6.12	1997	4.57
1966	4.74	1982	8.87	1998	9.13
1967	7.15	1983	9.02	1999	3.98
1968	7.15	1984	10.71	2000	5.17

***Source**: Natural Disaster Management Division.*

MITIGATION MEASURES

The objective of any mitigation measure might be to minimize the loss of lives and social, private and community assets and contribute to sustainable development. With this rationale, following mitigation measures are suggested based on the nature of flooding as a process, characteristics of rivers as a natural agent of hydrological cycle and the experience of various countries and control measures.

- As is the case of any geohazard, the best way to mitigate flood hazard is keeping away from vulnerable regions. That is, awareness on flood prone regions and protection of them from any developmental activities. However, unlike other geohazard prone regions, river courses and flood plains are the most sought after areas for any developmental activity including cultivation since the dawn of human civilization. With the advent of liberalization and evolution of global economy, optimal utilization of natural resources is required and it commences with establishment of settlements, infrastructures, laying of road and rail lines for transport, etc, on the very rivers and flood plains which, any mitigation measure would attempt to protect from developmental activity. Under this scenario, it is imperative to seek next best alternative - taking the risk of flooding with measures that can minimize the impact. Examples are galore in this regard, such as Vienna (structural measures in combination with flood regulation and control measures) and New Zealand (Seawalls, dune development and natural vegetation development) wherein effective and long lasting measures are implemented.
- Traditionally, structural controls such as embankments along river channels, construction of dams and reservoirs, development of bypass channels and construction of seawalls along coasts have been attempted to contain flooding impact. However, such structural measures invariably alter the natural flow conditions of rivers and storm surges resulting in habitat loss posing pressure on biological and ecological systems. Current practice of flood control includes integrated area development approach with minimal interference on natural flow conditions.
- As river channels and flood flow from them have the 100-200 year to 500 year cycles, the areas that have not been flooded for the past 200 - 500 years but are located within the reach of river

channels are to be identified as flood-prone areas and any development activity in those regions have to be restricted, if proper natural flow conditions could not be preserved during developmental activities. A case in this regard is the recent flood in Bihar, where extensive flooding occurred due to recapture of a Kosi river channel in an area through which the river had had its flood plain during hundreds of years back. Giving legal sanctity to this phenomenon of rivers, developed countries such as United Kingdom and USA have banned new constructions and restricted developmental activities in 200-500 years old flood plain regions and paleo river courses. Hence, identification of paleochannels of active/dry rivers and mapping their paleo flood plains followed by restriction of any alteration or preservation of natural flow/percolation conditions in those areas should be one of the prerequisites before taking up any major developmental activity.

- When left undisturbed, a floodplain provides storage area for floodwaters helping to reduce the height and flow of flooding. Floodplains also provide habitat for a diverse array of plants and animals, control erosion, filter runoff and recharge groundwater. Particularly important is the fact that when there are no buildings in a floodplain, damage by flooding to human life and property is greatly diminished.
- Mapping flood prone areas and assigning vulnerability index to different parts of the areas based on historical records of flooding, recurrence, period and the extent of flooding are essential. Kale (1997) suggested that flood hazard mapping is a prerequisite for minimizing life and property loss due to flooding.
- Geological Survey of India studies the shape and material of the land getting inundated and generates data on area, shape, slope, infiltration and permeability of soil of the basin, drainage pattern, landform and longitudinal and cross profiles of the channels and generates flood hazard maps indicating Prohibitive, Restricted, Cautionary and Flood Free Zones. Any developmental activity should take into consideration of these different levels of vulnerability in addition to local conditions and accordingly preventive measure should be made.
- In addition to hydrological and meteorological conditions, geological and geographic details, demographic details should also be taken into consideration in flood vulnerability indexes.

- As urbanization and developmental activities such as establishment of commercial, industrial and recreational activities invariably reduce the area of percolation, enhancing surface runoff, establishment of structures such as construction of rainwater harvesting, percolation ponds and check dams that promote infiltration commensurate with the area and amount of percolation thwarted by the developmental activities within that area should be made as a mandatory condition.
- In the coastal areas, the tide levels and the land characteristics will determine the submergence areas. Flood hazard mapping will give the proper indication of water flow during floods. Based on historical flood levels and changing geomorphic and landuse characteristics, adequate measures such as establishment of artificial levees - either as construction or natural vegetation barrier or a combination of both should be made in these areas.
- Conservation and reclamation efforts should be made not only in vulnerable downstream regions, but also in upstream regions. Jain and Sinha (2003), based on their study of Baghmati River, commented that flood management of riverine regions should concentrate on headlands where run off generation and distribution takes place instead of downstream of main channel alone.
- Even in a developed country such as Canada, according to Catto and Hickman (2005) identification of floods and other natural hazards, assessments of their impacts and suggestions for adaptation have proceeded in a somewhat piecemeal fashion. Fragmentation of databases, loss of records, institutional, community and individual memories, limited financial and personnel resources and difficulties with effective dissemination of the available information hamper efforts to minimize flood impact. These authors have suggested that an integrated approach, encompassing mapping, analysis, socio-economic research and vulnerability assessment is vital.
- Scientific land use planning and stringent control measures would greatly reduce flooding. While this measure is successfully implemented in developed countries such as United Kingdom, United States of America and Europe, lack of such stringent measures thwarts effective flood control in developing countries.

- In addition to stringent landuse control, flood forecasting and structural flood control measures, many countries resort to restoration of wetlands along river courses and flood plains with significant success in diminishing flood damage. By restoration, increase in forested lands and augmentation of infiltration are achieved, in addition to shifting people and infrastructure away from vulnerable areas. Similar "positive control" measures should be implemented and given priority.
- Controlling the flood flow and linking the rivers would be an efficient proposition to contain flood impact. This measure would be more appropriate for a country such as India wherein the geographic diversity is so extreme that while one part of the country reels under water shortage and drought, cyclones and floods simultaneously thrash other part. Effective linking of rivers and diversion of floodwaters to other river basins would serve well both for the purposes of flood management as well as meeting the demands of poorly drained river basins. However, this measure is prought with the danger of homogenization of river water and may act as conduits of endemics.
- Use of Geographic Information System (GIS) for flood hazard zonation by collating various parameters such as rainfall, distribution of rainfall, watershed characteristics, drainage density, morphometry, basin area, geomorphology, lithology, relief, slope, soil, channel characteristics, geographic and demographic characteristics of urban and industrial centers and land use and attempting reclamation and conservation measures to reduce flood hazard are suggested by many studies. Based on multi-parameter study of Newfoundland of Canada, Catto and Hickman (2005) have shown that flooding events are not statistically associated with overall wet years and also that total annual precipitation increases do not directly imply increased flood risk. Anomalously dry years are not necessarily associated with reduced risk of flooding and flood events are not directly associated with heavy precipitation following dry conditions
- As the hydrological response of a river system may vary for a given amount of rainfall (Jain and Sinha, 2003) integration of related parameters is essential towards effective flood management practices. Such systematic studies have to be

conducted at micro-watershed level scale and appropriate mitigation measures have to be implemented.

- The website of Natural Disaster Management Division lists that flood Management in India commenced with the Five Year Plans, particularly with the launching of National Program of Flood Management in 1954. During the last 48 years, different structural and non-structural methods of flood protection have been adopted in various states depending on the nature of the problem and local conditions. Structural measures include establishment of storage reservoirs, flood embankments, drainage channels, anti-erosion works, channel improvement works, detention basins etc. Non-structural measures include flood forecasting, flood plain zoning, flood proofing, disaster preparedness etc. The flood management measures undertaken so far have provided reasonable degree of protection to an area of 15.81 million hectares through out India.
- The website also states that flood management is implemented through community based programs such as sedimentation clearance, reforestation program and dike and flood wall construction. Similar works and refurbishment of natural water bodies are undertaken on a larger scale through community participation under the recently enacted law of Assured employment for rural people.

CONCLUSIONS

From the foregoing review of information on flooding characteristics, vulnerable regions and mitigation measures, it is clear that, although flooding is a natural and recurrent event of riverine process and part of natural hydrological system, transformation of this normal process into a disaster occurs only when overzealous human population occupies the land of the river. Hence, restoration of land belonging to natural flooding process wherever possible, implementation of positive control measures for minimizing flooding impact are suggested as effective mitigation measures. It is also demonstrated through this review that unlike other hazards, flooding could be managed through proper understanding, systematic study and positive approach.

References

Arya, S., Karanth, A. and Agarwal, A., 2005 Hazards, disasters and your community: A primer for parliamentarians. Government of India. 62p.

Catto, N. and Hickman, H., 2005 Flooding hazard identification, mapping and vulnerability assessment in Newfoundland, Canada. Geophy.Res.Abst. vol. 7. p.08542.

Government of India., 2004 Disaster management in India - A status report. National disaster management division. 88p.

Jain, V and Sinha, R., 2003 Evaluation of geomorphic control on flood hazard through geomorphic instantaneous unit hydrograph. Curr.Sci. v.85. pp.1596-1600.

Kale, V.S. 1997 Flood studies in India: A brief review. Jour.Geol.Soc.India. v.49. pp.359-370.

Parker, D.J., 2000 Floods. Taylor and Francis. 431p.

Sinha, R and Jain, V., 1998 Flood hazards of north Bihar rivers, Indo-Gangetic plains. *In:* Kale, V.S. (Ed.) Flood Studies in India. Mem.Geol.Soc.India. v.41. pp.27-52.

United Nations., 1976 Pre-disaster physical planning of Human settlement. Guidelines for disaster prevention. 35p.

World Meteorological Organization., 2005 Economic aspects of integrated flood management. 69p.

www.cwc.nic.in.

www.dartmouth.edu/~floods/archieveatlas/.

www.portal.gsi.gov.in/portal/.

❑❑❑

14

Environmental Policies of India and Mitigation of Geohazards in Coastal Areas

M. JAYANTHI
Central Institute of Brackishwater Aquaculture,
75, Santhome High road, R.A.Puram, Chennai - 620 028
E-mail: <jayanthivenki@rediffmail.com>

Many types of geological hazards namely, tsunami, earthquakes, tsunami, floods and drought affect coastal areas like any other place on the Earth. As coastal regions are thickly populated and have abundant urban, industrial and recreational establishments, impacts of geohazards are always severe along coastal regions. Compliance to coastal zone regulations, with protection of sand dunes and mangroves will be the best mechanism to minimize the damages from such hazards. In future, climate change may result in more hazards such as erratic rainfall and increase the frequency of droughts, floods and storm surges, etc. In this context, proposed coastal management zone for the coastal protection, disaster mitigation and management tools in coastal areas are examined in this chapter. It is suggested that extensive use of GIS and development of Geodatabase for the risk prone areas will serve as efficient tool for planners and policy makers to deal with geological hazards. In addition, it is felt that more efforts are needed in legal and institutional framework for integrated disaster management planning on local as well as national level.

INTRODUCTION

Geologic hazards are natural geological processes that inflict severe damage in areas of their operation if the area is utilized by anthropogenic activity in terms of habitation, industrial, recreational or commercial establishments. Many a times, natural geological processes that operated uninhibited for millions of years turn into geological hazards due to human intervention. These losses - lives and economic are not evenly distributed and are more prevalent in the developing countries due to higher population concentration and low level of economic growth. United Nations had realized the need of reducing these losses due to natural disasters and declared the last decade (1991-2000) as the International Decade for Natural Disaster Reduction. Objective of this proclamation was to reduce, through concerted efforts, the loss of life, property damage and socio-economic disruption caused by the natural disasters particularly in the developing countries (Tewar, 2000).

Major impacts of geohazards in coastal regions include heavy loss of life, settlements and infrastructure, agricultural lands, destruction of coastal ecosystems, adverse changes in ground water quality, deforestation and contamination of water sources. Climate change has also been interpreted as one of the reason for many of the geological hazards. However, this climate change is the very creation of intensive industrialization, urbanization and extensive use of fossil fuels. Given cognizance to the fact that the coastal regions are one of the very sensitive ecosystems on earth and also to their role in providing sustenance to human population and economic activity, this chapter presents an account on impacts and mitigation measures of geohazards along coastal regions.

COASTAL HAZARDS AND CLIMATE CHANGE

Owing to their low topography, susceptibility to easy inundation, erosion and sedimentation, coastal areas experience higher intensities of impacts from geohazards than other regions of the Earth. As typically coastal areas are thickly populated and utilized for varieties of commercial, industrial, and recreational activities, the damage caused by these higher intensities of impacts result in catastrophes even during slightest change in existing equilibrium of coastal geological processes. Owing to these reasons, scientists classified many coastal areas including 7500 km coastline of Indian sub-continent and adjoining regions as hazard prone zones. This designation has led to an increasing attention on the coastal risks and attempts to understand and mitigate them. When climate and man made pressures cause partial or total imbalance of the coastal system, it could result in the erosion of shores and habitat, increased salinity of estuaries and fresh water aquifers, altered tidal ranges in rivers and bays, changes in sediment and nutrient transport, increased coastal flooding and in turn affect the vulnerability of the coastal population (Bergkam and Orlando, 1999). Current climate change scenario predicts an increase of 2°C and a rise in sea level of approximately 1.5m within the coming half-century (IPCC 1996). Water temperature can have direct brunt on spawning and survival of larvae and juveniles and thereby on food availability. Accelerated rates in sea level rise will likely result in shifts in species composition, a reduction in wetland productivity and function (Warren and Niering, 1993) and reduce the fresh water supply available to coastal wetlands due to salt water intrusion.

Climate change may affect fisheries and aquaculture directly by altering the availability of fish and other consumable species and indirectly by changing the price of fish products and inputs. The impacts of sea level rise would cause reduction of land available for agriculture and aquaculture, loss of coastal eco systems, salt-water intrusion and shift in species abundance and distribution. Increase in frequency of cyclones or storm surges on account of climatic cycle change will result in inland flooding, salinity changes and introduction of disease or predators into newer ecological niches. As a result of climate change, drought would creep in and it will bring the changes in salinity and river flows and lead to the water shortages. Sea surface temperature changes will increase the occurrence of harmful algal blooms, disease outbreaks and growing seasons in aquaculture and decrease the dissolved oxygen in water bodies.

MITIGATION MEASURES FOR COASTAL GEOHAZARDS

Various measures are taken by the governmental agencies at all levels to prevent heavy loss of life and property during the crisis. Governmental agencies follow a system to mitigate geohazards and manage impacts. The system includes many components namely, identification and planning, mitigation, preparedness, response and recovery.

Owing to the occurrence of loose soils and sediments, the impact of earthquakes would be multifold in coastal areas. Hence, strict vigil and monitoring of existing structures, location of infrastructure projects only in areas safe from earthquake hazards and permitting construction works with earthquake resistant structures are to be followed. Furthermore, the enforcement of no-construction in areas located 500m from highest high tide line should be made without exceptions.

Although rock falls and mass wasting are rare along east coast of India, west coast poses a peculiar problem in areas wherein cliff coast is present. These areas are being undermined by relentless work of waves that get aggravated during seasonal cyclones and storms resulting in damage to coastal structures and unstable cliffs. Suitable stabilization efforts have to be made after studying the areas and applying site and measure specific remedies.

Sand dunes prevented the killer tsunami of December 2004 from ravaging areas beyond 132 meters from the sea, where there were none the devastating waves demolished structures as far as 862 m from the coast. The lesson learned from the tsunami clearly indicated that the losses must have been much lesser if our coastline is protected. Mangrove forests have acted as bio-shields against tsunami and saved many villages adjacent to this forest. As a general measure, preservation of sand dunes, conservation of mangrove ecosystem, identifying tsunami prone areas based on topography and other geological parameters, establishing green belt barrier in tsunami prone areas, enforcement and strict compliance of Coastal Regulation Zone norms and development of coastal hazard policy and its implementation would greatly reduce the potential impact of tsunami.

Flood situation in India is ubiquitous and arises almost every year and hence it is important to initiate activities for better management practices to contain its impacts. The Flood Forecasting Network of Central Water Commission (CWC) sends information to all the major flood prone inter-State river basins in the country. Information from satellites is used for mapping and monitoring flood prone areas. Apart from an effective disaster response system, it is important to have a good flood prevention and mitigation system to reduce the vulnerability of urban centres and commercial establishments. State governments have come forward to take up mitigation program like construction of raised platforms, embankments, and flood walls along various flood prone rivers, etc. Despite these efforts, floods continue to be a menace primarily because of the huge quantum of silt, which has raised the bed level in many rivers. Development of proper drainage networks in urban centers, clearance of paleo- networks of natural drainages and streams, periodic maintenance and frequent monitoring and maintenance of sludge and natural drainage systems during monsoon seasons, reduction of surface run-off and establishment of natural percolation of meteoric water into groundwater systems would greatly help reduce flood hazard. Overflowing rivers and storm surges cause coastal flooding. Specific efforts to contain this menace through establishment of bio-shields and structural measures are necessary in these regions.

It is an irony that India is a specific country which experiences flooding in one part and simultaneously drought in another part owing to the vagaries of nature and also due to geomorphology. As drought is a hazard that slowly develops in an area, better monitoring system would help reduce its impact. It can be managed to some extent with prior planning and preventive measures. The monitoring the pattern of rainfall and predicting the drought will help the government organizations to draw disaster management plan. Desalination of sea water for irrigation and/or domestic consumption, analysis rainfall pattern and conservation of water usage, land use planning, cultivation of short duration crops, adoption rainwater harvesting techniques, imposing restrictions on water usage are few of the measures that could avoid and or minimize drought conditions.

COASTAL ENVIRONMENTAL GUIDELINES

The Indian coastal zone is governed by several legislations that regulate development activities including urbanization, industrial and commercial activity and other coastal infrastructure. Few of these legislations have an explicit mandate to protect the coastal ecology and natural resources of the region. One such protective legislation is the notification of "Coastal Zone Regulation Act" in the year 1991 using the provisions of the Environment (Protection) Act, 1986 and the Environment (Protection) Rules, 1986. Through such a notification, the coastline of the country has been identified as an ecologically sensitive area, where development activities are regulated. The Coastal Zone Notification Regulation 1991, recognized four categories of coast lines namely CRZ I, CRZ II, CRZ III and CRZ IV and accordingly regulations and controls are applied to coastal stretches of seas, bays, estuaries, creeks, rivers and backwaters which are influenced by tidal action. The intertidal area along with a 500m zone from the high tide line is identified as a zone where development is restricted or prohibited, though exemptions are granted for permitted activities, which require waterfront or seafront access. No new constructions were permitted up to 500m in CRZ I and upto 200 m in CRZ III. Compliance to Coastal Regulation Zone (CRZ) norms will preserve our coastal areas and its ecosystem. CRZ compliance could have saved more lives during the 2004 Tsunami. CRZ notification was not adopted properly and many new constructions along the coast have come up at the cost of coastal vegetation.

PROPOSED COASTAL MANAGEMENT ZONE (CMZ) NOTIFICATION

The Committee headed by Dr.M.S.Swaminathan has evolved 12 basic guiding principles to govern future decisions on coastal zone management. The Committee has recommended reclassification of the coastal zone into four categories as outlined below:

Coastal Management Zone-I : It consists of areas designated as ecologically sensitive such as mangroves, coral reefs, sand dunes, inland tide/water bodies such as estuaries, lakes, lagoons, creeks and straits, mudflats, marine parks and sanctuaries, coastal forests, coastal fresh water lakes, salt marshes, turtle nesting grounds, horse shoe crab habitats, seagrass beds, sea weed beds and nesting grounds of migratory birds.

Coastal Management Zone-II : It consists of areas identified as "areas of particular concern" such as economically important areas, high population areas and culturally/strategically important areas. The administrative boundaries of these areas would be boundaries of CMZ-II.

Coastal Management Zone-III : It consists of all other open areas including the coastal seas but excluding those areas classified as CMZ-I, CMZ-II and CMZ -IV.

Coastal Management Zone-IV : It consists of islands of the Andaman and Nicobar and Lakshadweep.

The new CMZ will include an area of the sea to form territorial limits (12 nautical miles), including its seabed to the administrative boundaries or the biological boundaries demarcated on the landward side of the seacoast. The zone will also include the inland tidal water bodies influenced by tidal action and the land area along such water bodies. This area should be taken up for an integrated, cohesive, multi-disciplinary and multi-sectoral coastal area management and regulatory system.

The protection and sustainable development of the marine and coastal environment and resources should be in conformity with international law and conventions of United Nations. The Draft National Environment Policy of India also contains useful guidelines. It specifies that every effort should be made towards ensuring an Integrated Marine and Coastal Area Management (IMCAM) as prescribed in the 1995 Jakarta Mandate under the 1992 Convention on Biological Diversity.

TOOLS FOR COASTAL AREA DISASTER MANAGEMENT

Environmentally effective coastal zone management depends upon accurate and comprehensive scientific data, based on which the policy decisions are made. The existing information on these areas is patchy in coverage and lacks consistency. Further, a large time gap exists in many cases between data compilation and publication (Ramkumar et al. 2008). Remote sensing technique has emerged as an appropriate tool in providing coastal area spatial information because of its systematic approach in data acquisition through satellites, affordability, high level of precision and possibility of time series data. Satellite imageries are increasingly being utilized as important data sources for many studies including coastal zone mapping and management. Availability of repetitive, synoptic and multi-spectral data from various satellite platforms, viz. LANDSAT, SPOT, IRS, NOAA, GOES and IKONOS series have helped to generate information on varied aspects of the coastal and marine environments. Ocean color data from OCANSAT I OCM, SeaWiFS and MODIS, provide information on biological aspects useful for fisheries and coastal ecosystems. In India, satellite based information is being used for generating inventory on coastal habitats, landforms, coastal land use and shoreline condition, assessment of status of protected areas such as mangroves and coral reefs. Recent advances in remote sensing and geographic information have led to the development of hyperspectral sensors that have capacity to accurately delineate areas (spatial resolution) and characteristics of areas (spectral resolution). There are many applications which can take advantage of hyperspectral remote sensing are site selection, Environment Impact Assessment (EIA), geology, ecology and coastal water quality. These capabilities could be utilized for monitoring and evaluating critical parameters required for coastal zone and disaster management.

A large database with georeferenced information will serve as basic input for planning the preventive and management schemes to deal with geological hazards. To derive such database, Geographical Information System (GIS) will be an indispensable tool because of its capability to relate different forms of layers and derive conclusion by various analytical capability. GIS is extensively used to assemble information from different types of data, maps and digital elevation models. Census data and other relevant statistical abstract about the available infrastructure facilities could also be used to make the hazard map more useful.

Development of Integrated Disaster Management Plan should involve issue identification and assessment, public education, stakeholder consultation, prioritizing issues, formulation and adoption of management plans and capacity building among different stakeholders. For efficient compliance and enforcement, the local communities and the government should manage the resources jointly.

When an emergency management team/organization has decided to prepare a disaster plan, one of the first steps required is to identify possible hazards, assets at risk and the location of various support facilities. After identifying the geographic area, the different types of hazards that threaten a specific area need to be identified based on past experiences and the scientific body of knowledge for predicting threats to a specific area.

The information has to be collected on a local area basis and stored in GIS where the data would undergo further analyses, modelling and display to help identify if patterns exist as to where and when disasters repeat themselves and which areas are more at risk than others. When the types of hazards likely to impact a given area have been accessed and understood, the areas of most vulnerability can be easily identified from maps. Master plans can be formulated to address these areas of vulnerability. Ready availability of database with thematic layers in GIS environment could help make decisions at appropriate scale be it village or block or district or state or national levels.

Mitigation measures are usually left to the local, state or federal authorities depending on the magnitude and area of impact. Whilst many authorities in developed nations are now being proactive in taking measures to reduce the impact of disasters, many developing nations are waiting until after an extreme event has occurred before addressing the issue. In either case, GIS analysis is being used to assist authorities identify areas where mitigation effort should be concentrated. Within each zone, the number and type of each support facility able to assist in responding to a disaster needs to be delineated on maps and communicated to all concerned authorities and general public.

CONCLUSIONS

Geological hazards can be managed to some extent with prior planning and preparedness. Land use zoning and allotment of buffer zones, regulations including licensing for permitted activities and non-regulatory mechanisms will help conserving the coastal regions. Participatory approach, continuous voluntary monitoring, codes of practices of each activity are the components that should be designed and fully integrated into the scheme of development and management. Attention must be paid to legal and institutional frameworks that support integrated management planning on local to national scales. There is a great need for regionwise geological database, to be made available from village level for disaster management plan. The vulnerability assessment and awareness will help the planners and policy makers to take appropriate actions in minimizing losses.

References

Bergkam, G. and Orlando, B., 1999 Wetlands and climate change background paper from ICUN, www.ramsar.org.

Intergovernmental Panel on Climate Change., 1996 Climate Change-Impacts, adaptations and mitigation of climate change: scientific technical analysis. Contribution of Working Group II to the second Assessment Report of the IPCC. Cambridge University Press, Cambridge.

Ramkumar, M., Anbarasu, K., Sathish, G., Suresh, R. and Venkateswaran, S., 2008 Hyperspectral Remote Sensing of Coastal Sedimentary Terrains: A solution in sight for faster, reliable and cost effective Environmental Impact Assessment. *In:* Rajendran, S., Aravindan, S. and Jayavel Rajakumar, T., (Eds.) Hyper Spectral remote sensing and database management system, New India Publishing Agency, New Delhi. pp.121- 131.

Tewari R.P., 2000 Earthquake hazards and mitigation in India with special reference to North Eastern Region. ENVIS Bull. v.8. pp.15-22.

Warren, R.S. and Niering, N.A., 1993 Vegetation change on a northeast tidal marsh: Interaction of sea level rise and marsh accretion. Ecology. v.74. pp.96-103.

❑❑❑

15

Coastal Erosion in Dakshina Kannada and Udupi Districts, Karnataka Implications on Effective Coastal Zone Management

AVINASH KUMAR & K.S.JAYAPPA
Department of Marine Geology, Mangalore University,
Mangalagangotri - 574 199.
E-mail: <kumaravinash13@rediffmail.com>

This chapter documents shoreline changes of the Dakshina Kannada and Udupi districts of Karnataka State for the period between 1967 and 2005 studied by comparing the SOI topographic maps with IRS-1C/1D LISS-III satellite imageries through GIS techniques. Examination of coastal geomorphology, sediment flux and effectiveness of seawalls had indicated that the 130km long coastline of Dakshina Kannada and Udupi districts is essentially stable, but for seasonal variations. It has been estimated that, out of 130km, about 55km long coast is vulnerable to sea erosion, of which, 25km is covered by seawalls and covering another 10km length underway. It is revealed through field examination that protection of coastline through these measures is satisfactory only along few patches while other stretches of seawalls simply shift the zone of erosion to adjacent coastline owing to many reasons including quality of construction, unscientific positioning of seawalls, etc. It is also inferred that same is the case of many other coastlines elsewhere including Kerala coast. These experiences warrant scientific design and implementation of coastline protection measures.

INTRODUCTION

The 130km long coastal zone of Dakshina Kannada and Udupi districts is one of the densely populated coastal zones of India. The coastline is almost straight, regular and oriented towards N22°W, N12°W and N6°W directions at south of Mangalore, north of Mangalore and north of Malpe respectively. Barrier beaches and spits, ancient beach ridges, marshes, swamps and tidal flats characterize this zone. Natural features and man-made structures interrupt the shoreline. The natural features include estuarine mouths of Netravati-Gurpur rivers near Mangalore, Mulki-Pavanje rivers near Mulki, Udyavara river near Malpe, Sita-Swarna rivers near Barkur, Kollur-Chakra-Haladi rivers near Kundapur and rock exposures near Someshwar, Surathkal, Mulur, Kapu and north of Malpe (Fig. 1). The man-made structures include one major and four minor ports; breakwaters at the Netravati-Gurpur estuary, New Mangalore Port and Malpe harbor and a number of seawalls placed at various places along the coastline (Fig. 2). All these structures/features bring about local irregularity and contribute to erosion and accretion along the shoreline to some extent. Apart from these, source of beach material, coastal processes like wave refraction, longshore current, littoral drift etc. are also responsible for shoreline change and beach morphology. Erosion of beaches on the one hand and siltation of navigable estuaries and harbors on the other are some of the major problems of this coast. Encroachment into active zone of beaches is yet another problem. To address these problems scientifically through understanding ongoing processes, many institutions and organizations including the Mangalore University, College of Fisheries, the Marine Wing of the Geological Survey of India, Mangalore; National Institute of Technology, Karnataka and Karnataka Engineering Research Station, K.R. Sagar have conducted systematic studies. This chapter reviews the results obtained by these organizations and documents the present scenario regarding beach erosion, shoreline changes, coastal defense work and effects of seawalls on beach morphology etc.

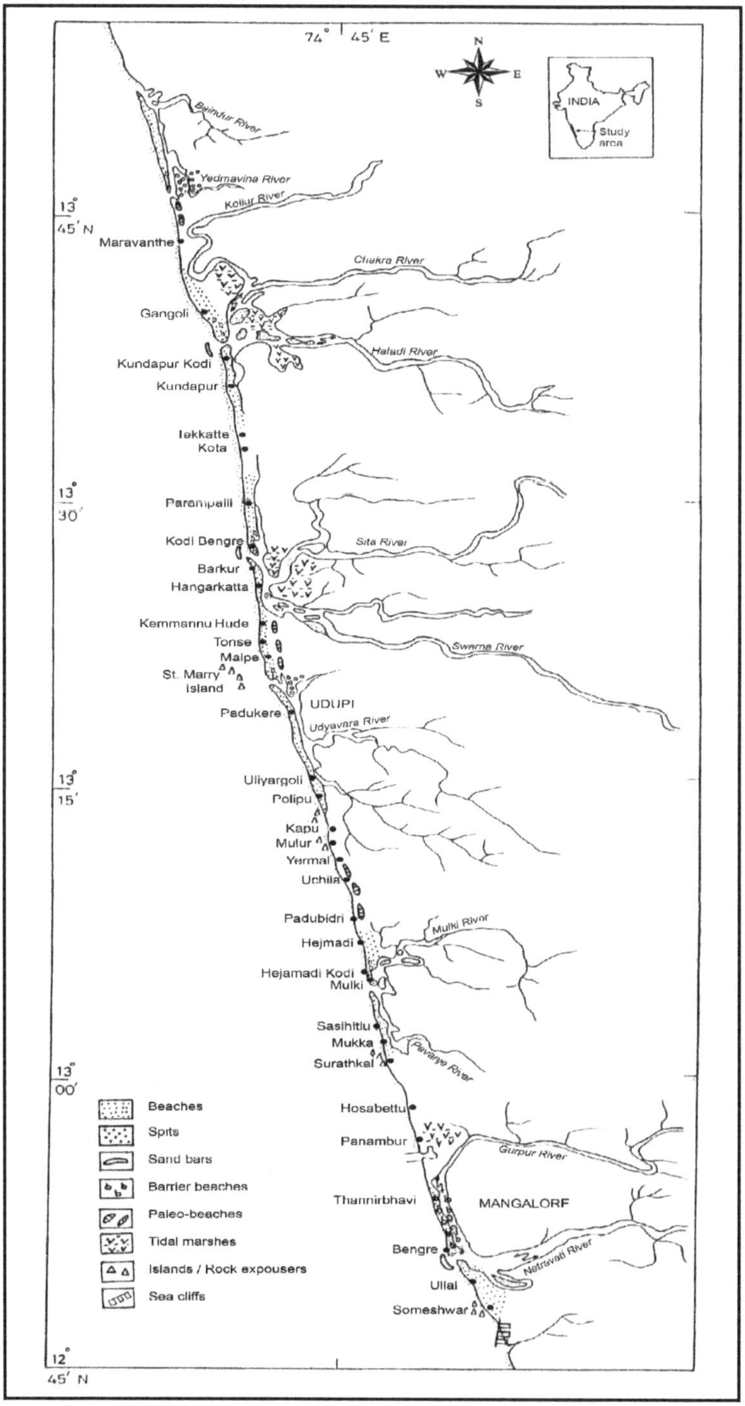

Figure 1. Coastal geomorphic features of the study area

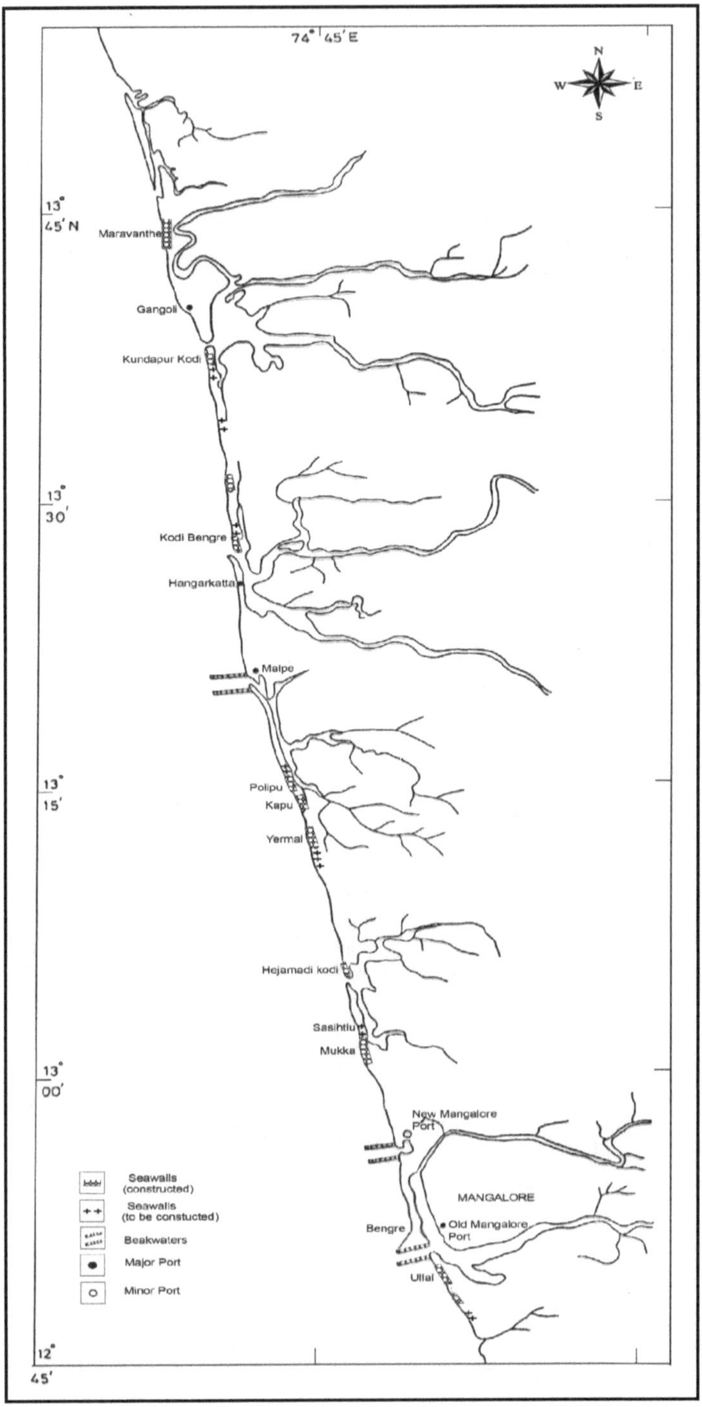

Figure 2. Locales of man-made structures along coast

DYNAMICS OF BEACH EROSION

Gole (1963) was the first to use fluorescent tracers to understand the movement of sediments at Old Mangalore Port in order to select a suitable site for an all-weather port in the nearby region of Mangalore. His study revealed that the movement of the bed material of the Netravati River was towards south. Based on grain-size statistics of continental shelf off Mangalore, Narayana and Pandarinath (1991) recognized transport of sediments towards onshore in regions of reduced fluvial influence and bidirectional transport (onshore and offshore) in regions of fluvial influence. Suryaprakasa Rao and Kassim (1970) studied beach profiles at Surathkal and observed that about 16m width of the beach was exposed to erosion during the SW monsoon and 19 m^3 of sand got eroded per linear meter of beach. Wave refraction studies of this coast indicated northerly-directed littoral transport for the waves approaching from SW and WSW directions (Reddy and Varadachari, 1972). John and Cheryan (1974) during their geomorphological studies of the Netravati estuary noticed an unique phenomenon of development of sand spits on either side of the Netravati-Gurpur estuarine mouth. A study by Reddy *et al.* (1977) on coastal erosion revealed considerable erosion on the northern side and to a certain extent in the middle portion of the Bengre beach. During their beach profile and rip current studies near Someshwar, Hariharan *et al.* (1978) observed accretion of beach from January to May and erosion of it from June to September.

Reddy and Hariharan (1979) reported considerable erosion at the mouth of the Netravati-Gurpur estuary near Bengre and also along the seaward side of the fishing village. They have estimated erosion of about a km long coastline along northern part of Bengre beach during a period between 1968 and 1978. Geomorphological studies conducted along the Kundapur coast have revealed that the area north of the Gangoli village was erosion prone and the Haladi River mouth was almost silted up. Of the two wave patterns (Northwest and Southwest) observed in this area, the waves from the northwest are found to be dominant, generating littoral drift southward resulting in growth of northern spit and erosion of southern one (Shrivastava, 1976). Morphological features that are long-term indicators also have been used as a tool in the study area to understand the processes responsible for beach morphological changes and net longshore drift (Murty, 1977).

Stability of Ullal beach except near the estuarine mouth, erosion of the Bengre beach and accretion to certain extent in beach at Thannirbhavi were observed by Reddy *et al.* (1982) based on beach characteristics, wave and littoral currents. Analysis of fluorescent tracer dispersion pattern on beach sand movement at Kundapur (Venkatesh, 1982) confirmed the interpretations of Shrivastava (1976) that the predominant vector of sand movement is alongshore in a southerly direction, signifying prevalence of similar processes for about a decade. Pore water pressure at coast-parallel river bends and blind tributaries is attributed to severe beach erosion in the study area (Ravindra and Krishna Rao, 1987). Morphological studies conducted on monthly time scales along the Ullal, Bengre and Thannirbhavi beaches and seasonal time scales along the Panambur and Surathkal beaches reveal that the beaches experience a phase of major erosion during the SW monsoon and they regain the lost sand during the fair weather season (Jayappa, 1988). Seasonal profile surveys conducted along the Panambur beach on either side of approach channel to the New Mangalore Port indicated that the beach north of the approach channel is wider and gentler than the region south of approach channel. The average width in the northern part is about 100m with a slope of 10° whereas on the southern side it is about 70m with a slope of 18°. From all these observations, net southerly littoral transport of beach sediment in this region was interpreted and the same has been inferred by tracer experiments conducted by Jayappa (1996).

Seasonal profiles of the Surathkal beach show maximum erosion of beach face during the monsoon (July), rapid accretion during the pre-monsoon period (March) and normal pace of accretion during post-monsoon period. The width of the beach is seen to be maximum (100-130m) during March and minimum (70-100m) during July. Slope of the foreshore is steep (32-45°) during the monsoon, whereas a moderately steep slope (25-30°) is seen to prevail during the pre-monsoon period. The volume changes per meter length of the beach over an annual cycle were estimated by using monthly profiles of Ullal, Bengre and Thannirbhavi beaches. These estimations indicated that about 115.8 m^3 of material per meter length of beach is added on to the Ullal beach and about 10 m^3 and 61.2 m^3 per meter length of beach is removed from the Bengre and Thannirbhavi beaches respectively. From these figures, it is clear that there is an overall addition of material (44.6 m^3/m) to the beaches when the profiles of all the three beaches are considered together (Jayappa, 1988). The interference of annual cyclic change in beach morphology is brought about by estuarine dynamics and man-made structures. This results in the accumulation of sediment on the upstream direction of breakwaters and erosion in the downstream direction (Jayappa and Subrahmanya, 1989). The study on the impact of coastal structures on the shoreline revealed that the erosional areas are migratory in nature in a few places and in a few other places it is due to anthropogenic causes (Subrahmanya et al. 1989). Nayak and Chavadi (1989) reported southerly transport along the coastal zone of northern Karnataka based on heavy mineral distribution. Kunte (1994) had drawn similar conclusion based on the length of drift sectors. He had attributed the development of large spits of this coastline to positive sediment balance. As most of the sediment is trapped in estuaries, the riverine contribution to the beaches/spits being minor, the surplus material comes from the erosion of shores and cliffs of the Uttara Kannada coast.

Kumar *et al.* (2006) reported coastal erosion and submergence of land at Ankola, Bhatkal, Malpe, Mulur, Mangalore, Honnavar, Maravanthe and Gokarna in Karnataka. The problem is relatively more severe in Dakshina Kannada and Udupi coasts, where about 40% of the total stretch is vulnerable to erosion during the SW monsoon. Shifting of river mouths near Kundapur (Kollur-Chakra-Haladi), Malpe (Udyavara) and Mangalore (Netravati-Gurpur) towards south also has been observed (Nayak *et al.* 1992). Shifting of the Gurpur River mouth was revealed by earlier studies (Subrahmanya and Jayappa, 1987; GSI, 1987). Study of dynamics of surf zone between Bhatkal and Ullal during 1989-90 by Chandramohan *et al.* (1994) revealed that the beaches along this segment of coastline are essentially stable but show seasonal fluctuations. They are being subjected to erosion during the SW monsoon but found to regain almost all the lost sediment by January or February.

Compilation of the review presented in preceding paragraphs state that since the last three decades, beach erosion in the Dakshina Kannada and Udupi districts has become highly problematic and as a result of which many studies have been conducted to understand the pattern and quantum of erosional process. Earlier, Ullal, Kotepura, Sasihitlu, Surathkal, Mukka, Hejamadi Kodi, Badagrama, Mulur, Kapu-Polipu, Padukere, Parampalli, Kundapur Kodi, Mannur, Gujjadi and Maravanthe were found to be erosion prone beaches (Fig. 3 and

Photograph 1 and 2). However, during recent years, it has been observed that erosion is not restricted to these beaches only, but it migrates to the adjoining beaches lying either towards the north or south. The migration of eroding sites could be due to the shifting of wave convergence zones and/or man-made structures. Sand and clay mining, construction of vented dams and decline in rainfall are the three main factors responsible for reduction in sediment supply and beach erosion. According to an estimate of the Minor Irrigation Department, State Government of Karnataka, about 55km length of this coastline is found to be vulnerable to sea erosion during the SW monsoon. After the monsoon, by the end of fair weather season, the beaches not only regain all the lost sediment, but also additional sediment in some areas. In general, the coastline in the study area is in a state of dynamic equilibrium except for areas near the river mouths that suffer permanent erosion due to natural shifting and migration of river mouths. Beaches near few seawalls also suffer erosion due to wave refraction and convergence.

Taking cues from these previous studies on dynamics of beach erosion and accretion and also the shifting nature of locales of active erosion/depositon due to natural and anthropogenic structures, it has become imperative to analyze the cumulative effect of these features as well as dynamics on a long-term basis in order to asses vulnerable regions and effectiveness of natural and man-made structures to contain erosion in the study area. Attempt in this regard and the results obtained are discussed in the following sections.

SHORELINE CHANGES

The nature and quantum of shoreline changes were estimated by comparing the Survey of India (SOI) topographic maps (1967 edition with 1:50,000 scale) with the imageries of Indian Remote Sensing Satellites IRS-1C/1D LISS-III obtained during January 1997 and December 2005. All the images were rectified (GCPs < 0.6 pixel) with reference to SOI topographic maps and False Colour Composites (FCC) were generated on 1:50,000 scale. Using ArcGIS software, the FCC's and the topomaps were digitized, superimposed and progradation/ recession of shoreline were measured to evaluate long-term (1967-97) and short-term (1997-2005) changes. Figure 3 depicts the results of this exercise and following observations are made from this figure and field survey.

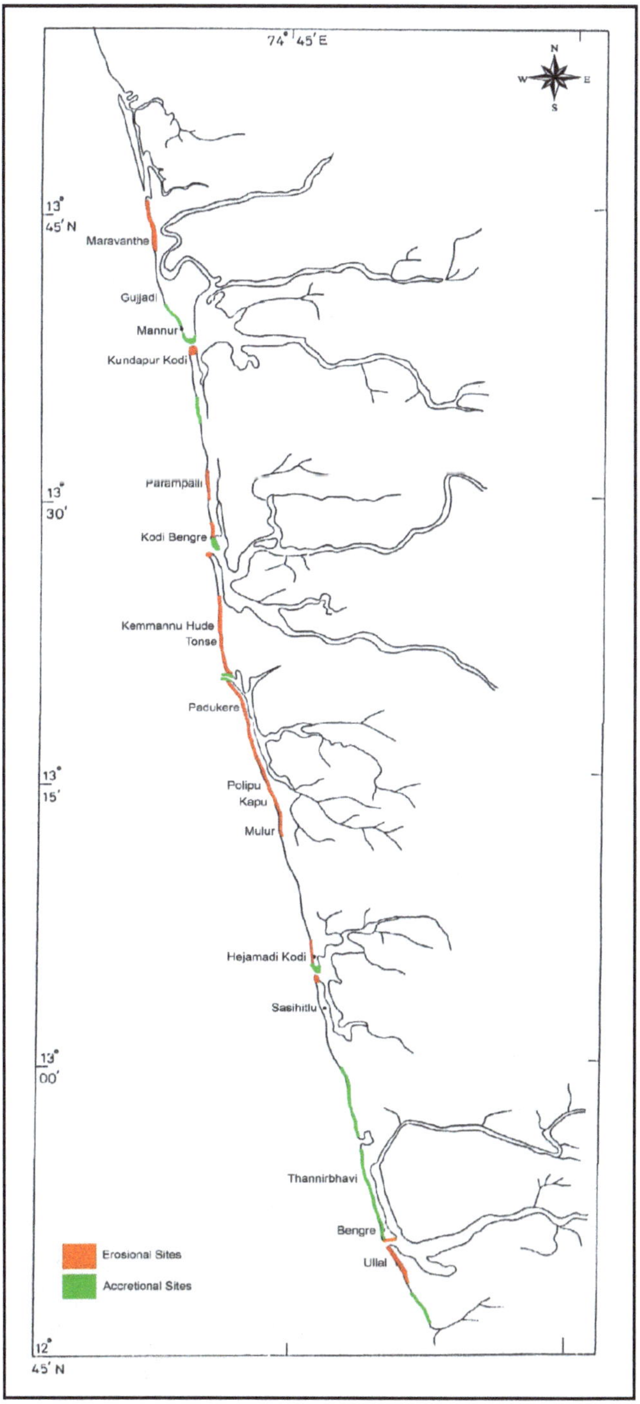

Figure 3. Location of erosional and accretional sites

Photograph 1. Severe damage of dwellings behind seawall at Ullal.

Photograph 2. Erosion of beach very close to NH 17 at Maravanthe.

Photograph 3. Shoreline with open beach at Thannirbhavi.

Photograph 4. Coastal defense using hard (armour rocks) and soft (Gabion bags) stabilization.

Photograph 5. Coastal erosion mitigation by proper construction of sea wall along Mukka. A vegetation line supports the wall from behind.

It is observed that the shoreline has prograded by 10 to 120m at some places and receded by 10-260m at other places. The beaches from Bengre to Thannirbhavi are open and maximum progradation is found here (Photograph 3). Ullal and Kotepura beaches are being subjected to severe erosion since the year 1997. Southern part of Someshwar beach has accreted during 2001-2005. Beaches from Talapadi to Someshwar have prograded between 1967 and 2001 but are subjected to recession during 2001-2005. Based on variations in width and slope of beach on either side of the approach channel to NMP and OMP, it is inferred that movement of sand from north to south is more than that from south to north. Construction of seawalls has resulted in shifting of erosion sites from one place to another, whereas, breakwaters have been acting as barriers for littoral drift.

The Sasihitlu spit has receded and shifted towards south but width of this spit had increased during 1967-97 period. Whereas, during 1997-2005, the length of this spit was reduced by ~280 m and its width also shows reduction. In general, the Hejamadi Kodi spit had grown towards south and SE directions from 1967 to 2005. The beaches from Padubidre to Yermal are stable whereas, Badagrama, Mulur, Kapu, Polipu and Uliyargoli to Kote beaches are being continuously subjected to erosion.

Dynamic changes were noticed at the Udyavara river mouth during 1967-97, but it is stable from 1997 onwards. Beaches between Malpe and Badanidiyur were erosional during 1967-97, and 1997 onwards accretion has been noticed. The Oddu Bengre spit is receding towards south, whereas, the Kodi Bengre spit has shifted northward and prograded towards sea. The Kodi spit is almost stable, except minor erosion at few places. No changes were noticed along the Gangoli spit during the 2001-2005.

Having observed dynamic nature of erosion/deposition sites on a short and long-term basis and the vulnerability of study area to continued onslaught of erosion, effectiveness of beach stabilization efforts is analyzed through field examination. This is essential to evolve a strategy for the betterment of coastal protection measure and also to apply effective measures elsewhere for mitigation of erosion hazard.

COASTAL DEFENSE WORK AND ITS EFFECT ON SHORELINE

Beach stabilization and or protection could be made through hard or soft stabilization. Hard stabilisation refers to establishment of seawalls, revetments, bulkheads and groins. Soft stabilisation refers primarily to beach replenishment with soft sediments (Pilkey and Wright III, 1988). Since the cost of soft stabilisation is tremendous, hard stabilisation, particularly seawall emplacement, is practiced in India including the coastline along the Dakshina Kannada and Udupi coastline (Photograph 4 and 5). However, whenever dredged material is available, it is used for beach nourishment in this region.

Work on provision of coastal protection structures such as construction of seawalls started in Karnataka during 1978-79. About 25km length of the coastline comprising Ullal, Bengre, Mukka, Hejamadi Kodi, Yermal, Kapu, Polipu, Kodi Bengre, Parampalli, Kundapur Kodi and Maravanthe beaches has been filled by single/ double layers of granite blocks (Fig. 2) in the form of protective wall at a cost of about Rs.15 crores. Construction of such walls along the beaches of Sasihitlu, Uchila, Uliyargoli and Tekkatte measuring about 10km is in different stages (Fig. 2) and the projected cost is Rs. 15 crores. It is also evident from the earlier studies that bigger the size of the structure, greater the damage on the coastal property. Such failures have also been reported from Kerala coast (Pillai, 1980). This is attributed to the channeling of coastal wave and current energy towards particular direction and or location immediately on construction of such structures. Experience from different coastal locations of the World suggests that coastal erosion could not be contained by mere construction of such walls or soft stabilization. A systematic study concerning oceanographic and sedimentary budgetary aspects including that of coastal geomorphology should be conducted and based on which, suitable shape and size of the protection structures have to be made; failing which, the "so called protection walls" would merely shift the problem elsewhere instead of minimizing or containing the problem.

The literature and field evidences available elsewhere indicate that most of the seawalls stabilize and protect the property present behind them. At the same time, the beach in front of the wall may become narrow and eventually disappear. The beach profile would become steep, and wave energy increases. The fact is that the protected property (resource) belongs to a few individuals relative to the number of people who use the beaches. Moreover, the cost of saving beach property by stabilization is very high. Often it is greater than the value of the property to be saved when long-range maintenance costs are considered (Kraus, 1988). The Bengre beach, which was vulnerable to sea erosion during the late 1970s and early 1980s, has become an accreting one after construction of a seawall in 1984 and attained maximum width subsequent to the construction of breakwaters in 1994. In contrast, the Kotepura spit started eroding in 1985 and continued the same trend until now and has become very narrow and steep (Jayappa and Vijaya Kumar, 2003; Avinash Kumar and Jayappa, *in press*). A study on influence of seawalls and breakwaters on the beaches in the study area has revealed that there is no balance between erosion and accretion trends in the engineered beaches unlike in open beaches (Jayappa *et al.* 2003). Seawalls properly located and constructed have performed satisfactorily, while those built in the active zone of the beach have either just shifted the problem to adjacent site or they themselves have been destroyed by huge waves (Dattatri, 1995 and 2000; Jayappa, 1998; Jayappa *et al.* 2003).

CONCLUSIONS

From the foregoing review, the following broad conclusions are made.

- In general, the beaches of Dakshina Kannada and Udupi districts are stable. About 55km length of this coastline is vulnerable to erosion, but it is a temporary phenomenon that takes place during the SW monsoon.
- Seawalls properly located and constructed perform satisfactorily while those built in the active zone of the beach have suffered damage.
- Longshore drift plays the primary role in the formation and growth of spits in this area, while river drift plays a secondary role. Longshore drift direction is either towards the north or the south depending upon the monsoonal season. Open beaches are in dynamic equilibrium and only certain vulnerable areas need to be mitigated.
- The authors are of the opinion that there is no permanent solution for coastal erosion but green walls are effective than seawalls. Leaving the shoreline as it is should be a better option, so that nature will maintain the dynamic equilibrium. Construction of seawalls may be taken up only at those places where it is very essential. However, their positioning, quality of construction, etc should be based on scientific design instead of blindly positioning them.
- Ports should maintain a few kilometers of beaches on either side by spending at least a part of their profit and dredged material should be used for only for beach nourishment.
- There should not be any developmental activity within the Coastal Regulation Zone. Although this regulation is in practice literally, strict implementation would reduce greately the negative impacts.
- Creation of digital database on coastline changes and establishing a monitoring scheme using remote sensing and Geographic Information System are required for effective coastal zone management.

References

Avinash Kumar and Jayappa K.S., 2008 Long and short-term shoreline changes along Mangalore coast, India. Inter.Jour.Env.Res. pp.(*in press*).

Chandramohan, P., Sanil Kumar, V., Nayak, B.U. and Raju, N.S.N., 1994 Surf zone dynamics along the south Karnataka coast between Bhatkal and Ullal, west coast of India. Ind.Jour.Mar.Sci. v.23. pp.189-194.

Dattatri, J., 1995 Report on coastal erosion and strategies for coastal protection in Dakshina Kannada. Technical report submitted to DANIDA. 241p.

Dattatri, J., 2000 Coastal erosion and protection along Karnataka coast. S.D.M.C.E.T., Dharwad & IGCP-367. Spl.Publ. v. 2. pp.97-100.

Geological Survey of India., 1987 Geological and geophysical evaluation of Old Mangalore Port area, Karnataka (SSAU-31, 32 & 33 cruises) – A geotechnical report. OME and MG wing Newslett. v.3. pp.21-22.

Gole, C. V., 1963 Fluorescent tracer study at Mangalore Port. Jour.Inter.Assoc.Hydr.Cong. v.1. pp.275-282.

Hariharan, V., Reddy, M.P.M. and Kurian, N.P., 1978 Littoral and rip currents and beach profiles off Someshwar. Ind.Geogr.Jour. v.53. pp. 14-20.

Jayappa, K. S., 1988 A study of beach morphology and nearshore geomorphology of the Mangalore area. Ph. D. thesis submitted to Mangalore University, Mangalore. (Unpublished).

Jayappa, K.S., 1996 Longshore sediment transport along the Mangalore coast, west coast of India. Ind.Jour.Mar.Sci. v.25. pp.157-159.

Jayappa, K.S., 1998 Coastal zone management with special reference to Karnataka. *In:* Hosetty, B.B. and Kumar, A., (Eds.), Environmental Impact Assessment and Management. Daya Publishing House, Delhi. pp.196-211.

Jayappa, K.S. and Subrahmanya, K.R., 1989 Beach morphological changes between Talapadi and Surathkal, Dakshina Kannada. Geol.Surv.India. Spl.Publ. v.24. pp.299-306.

Jayappa, K.S. and Vijaya Kumar, G.T., 2003 Coastal dynamics related to engineering structures and management issues: Two case studies from DK district, Karnataka. Jour.Ind.Assoc.Sediment. v.22. pp.233-241.

Jayappa, K.S., Vijaya Kumar, G.T. and Subrahmanya, K.R., 2003 Influence of coastal structures on the beaches of southern Kannada, India. Jour.Coast.Res. v.19. pp.389-408.

John, E.J. and Cheryan, K.P., 1974 Geomorphological studies of the estuary of the river Netravati near Mangalore. Proc.Coast.Engg.Conf. ASCE. II. pp.1304-1318.

Kraus, N.C., 1988 The effects of seawalls on the beach: An extended literature review. Jour.Coast.Res. v.4. pp.1-28.

Kumar, V.S., Pathak, K.C., Pednekar, P., Raju, N.S.N. and Gowthaman, R., 2006 Coastal processes along the Indian coastline. Curr.Sci. v.91. pp.530-536.

Kunte, P.D., 1994 Sediment transport along the Goa - north Karnataka coast, western India. Mar.Geol. v.118. pp.207-216.

Murty, P.S.N., 1977 Evolution of Netravati drainage, Karnataka state, south India. Ind.Jour.Earth Sci. v.4. pp.197-202.

Narayana, A.C. and Pandarinath, K., 1991 Sediment transport direction derived from grain-size statistics on the continental shelf off Mangalore, west coast of India. Jour.Geol.Soc.India. v.38. pp.293-298.

Nayak, G.N. and Chavadi, V.C., 1989 Distribution of heavy minerals in the beach sediments around Kali River, Karwar, west coast of India. Geol.Surv.India.Spl.Publ. v.24. pp.241-245.

Nayak, S., 1992 Coastal environment: a scientific report No. RSAM/SAC/COM/SN/11/92.

Pilkey, O.H. and Wright III, H.L., 1988 Seawalls versus beaches. Jour.Coast.Res. v.4. pp.41-64.

Pillai, N.A., 1980 Performance of coastal protection works along the coast of Kerala. Nat.Sem. on Protection and control of coastal erosion in India. National Institute of Oceanography, Goa. Abst.vol. pp. 36-37.

Ravindra, B.M. and Krishna Rao., 1987 Relation of coastal faults and river morphology to sea erosion in Dakshina Kannada, Karnataka. Jour.Geol.Soc.Ind. v.29. pp.424-432.

Reddy, M.P.M. and Hariharan, V., 1979 Problems of coastal erosion in India. Seafood Exp.Jour. v.11. pp.1-4.

Reddy, M.P.M., Hariharan, V. and Kurian, N.P., 1977 Coastal erosion at Bengre, a fishing village near Mangalore - A preliminary study. Mahasagar, Bull.Nat.Inst.Ocean. v.10. pp.97-102.

Reddy, M.P.M., Hariharan, V. and Kurian, N.P., 1982 Beach stability along the coast from Ullal to Thannirbhavi near Mangalore. Ind.Jour.Mar.Sci. v.11. pp.327-332.

Reddy, M.P.M. and Varadachari, V.V.R., 1972 Sediment movement in relation to wave refraction along the west coast of India. Proc.Ind.Geophys.Union. v.10. pp.169-191.

Shrivastava, P.C., 1976 A note on the geomorphological studies around Kundapur, west coast of India. Ind.Miner. v.30. pp.84-86.

Subrahmanya, K.R. and Jayappa, K.S., 1987 Origin of Netravati-Gurpur estuarine system. Proc.Nat.Sem. on Estuarine Management, Trivandrum. pp.70-72.

Subrahmanya, K.R., Jayappa, K.S. and Suresh G.C., 1989 Impact of coastal structures on Dakshina Kannada shoreline. III Nat.Conf. on Dock and Harbour Engg. Surathkal. v.2. pp.541-547.

Suryaprakasa Rao, P. and Kassim, K., 1970 Field study of Surathkal beach. Coast.Engg.Conf. Washington, D. C., II. pp.1365-1381.

Venkatesh, K.V., 1982 Study of beach sand movement by fluorescent tracer at Kundapur, South Canara district, Karnataka. Jour.Geol.Soc.India. v.23. pp.444-449.

16

Quicksand : A Lesser-Known Geohazard; But not a Lesser Evil

***V. THIRUKUMARAN & **Mu.RAMKUMAR**
*Department of Geology, Government Arts College (Men), Salem - 636 007.
**Department of Geology, Periyar University, Salem - 636 011.
E-mail: <mail_kumaran@yahoo.co.in>

This chapter sheds light on a geohazard namely quicksand. Unlike other geohazards, it waits for its victims to cause hazard. Though discounted as a myth and also as a monster problem in Hollywood film stories, phenomenon of quicksand is real and could inflict life threatening dangers. Analyses of genetic causes, associated properties and locales of its occurrences indicate that more deaths might have resulted due to panic and drowning in waters than by quicksand itself. Based on these inferences, potential mitigation measures are presented in this chapter.

INTRODUCTION

Quicksand, referred as *"pudhai manal"* in Tamil, is one of the lesser known dangerous natural phenomena that cause death of human and other unsuspecting animals. It is a geohazard of different sort. It is different in a sense that it never "goes" anywhere like tsunami or causes hazard like flood or landslide or any other geohazard; but it waits for hapless victims; that is., it is harmless until human beings or other fauna step into the region of quicksand. It is commonly prevalent in many river beds and other locales near water bodies located amidst dense forests. Many a deaths occur due to quicksand but go unnoticed owing to its "non catastrophic" nature. It is also a geohazard less understood pertaining to causes of formation, potential to cause damage and occurrence (www.mythbuster.com). This chapter is an attempt to review the existing literature on this geohazard for better understanding.

THE PHENOMENON

Quicksand is of fine sand or silt or clay and salt "floating" on water. Quicksand is the terminology used for a region which consists of liquefied sand and silt that float over water whose depth may vary upto many meters. The sand particles in such regions are interspersed with thin film of water, clay and organic matter that make a weak bond between grains. The sandy layers that float over water are in a semi-liquid state. Any alteration of existing equilibrium of bond through application of pressure by stepping on to the floating sand layer would cause collapsing resulting in drowning of the object/human/fauna. The real danger rests not on the collapse of the weak bond, but due to the depth of water body over which the floating sand rested (resulting in drowning and death), stickiness of the flocculated water (death in panic) and entering of sediment laden water into the respiratory system and its blockage (slow death due to exasperation).

GENESIS

Three types of quick sand are reported. First is, under normal circumstances, sand or silt particles do not float on water. Under circumstances of upward movement of water through a bed of sand, silt and clay admixture and with a proper amount of hydrostatic pressure generated by sub-aqueous springs, the sand grains become a sand-water suspension resulting in formation of quicksand layer. While the sand layer remains solid interspersed with water molecules in between, even a slightest disturbance could make the "floating sand" to sink. Quicksands have higher specific gravity of 1.156 (http://en.wikipedia.org/ wiki/Quicksand) than that of underlying water (1.00). The stability of the colloidal quicksand is increased by the presence of salt, increasing the likelihood of sand flocculation and the formation of high viscosity regions of sediment responsible for quicksand's "trapping" power. It's the difficulty of moving this dense sand that causes the problem (Anonymous, 1984; Khaldoun *et al.* 2005).

Second is that, in a still water body located amidst forest canopy or thick vegetation, fine sediments float; over which, fine detritus and leafy materials get deposited slowly. Over time, such layers of sand-decayed organic matter-silt and clay detritus form a layer that float over water owing to their buoyancy. This floating layer of sand is stable but for application of higher pressures such as stepping in by human beings or large animals.

These two types of quicksands are composed of sand, silt, clay, water and salt. When pressure is applied over quicksands, the clay changes from gel state to liquid state, resulting in collapse of the floating nature, leading to sinking of sand to the bottom and the water particles move up. As soon as the clay particles liquefy, the salt and the clay particles get glued to each other making the floating materials to be sticky and these lumps settle with sand (Ramon and Wessner, 1996; Leslie, 1990), the resultant material is the real "culprit" that traps the victims.

The third is dry quicksand which is loose sand whose bulk density is reduced by blowing air through it and which yields easily to weight or pressure (http://en.wikipedia.org/wiki/ Dry_quicksand; Lohse *et al.* 2004). It acts similar to regular quicksand, but contains no water and does not operate on the same principle. Until recently, its existence was doubted, and the reports of humans and complete caravans being lost in dry quicksand were considered to be folklore (www.Mythbusters.com; Leslie, 1990). In nature, dry quicksands may evolve from the sedimentation of very fine sand after it has been blown into the air and, if large enough, might be a threat to humans.

OCCURRENCE

Quicksand occurs in three different environments. First is the soils along banks of ephemeral rivers, more particularly along their inner curves (point bars) of rivers and at steep slopes of advancing sheet sands that occur as bed load. Second is marshy land adjoining perennial water bodies such as rivers, lakes, coastal lagoons and tidal flats. Third is hilly area with loose soil and shallow ground water. It appears from this list that, quicksand can form almost anywhere wherein unconsolidated sand, shallow water body, protection from destabilization agents such as waves and currents, etc. are available.

MITIGATION MEASURES

There seem to be a dearth of research material available on quicksand. Many a times it is discounted as more of a myth than of reality. As condition of formation of quicksand is not perennial, there is a paucity of conductance of systematic research on it. Most of the information available on it is the "horror stories" of getting out of quicksand from remote regions and Hollywood film stories (www.Mythbusters.com). Nevertheless, the following could be followed to avoid potential dangers of quicksand.

- Avoid the uncharted paths in regions listed in previous section.
- In the event of getting stuck, spread the area of coverage to increase buoyancy. Any panicky reaction inevitably gets the victims still deep into the quicksand, thereby increasing the chances of fatality.
- More often quicksands in river beds cause fatalities by drowning and entering of sand-laden water into the respiratory system. Avoidance of such suspicious wetland is the best solution.
- In marshy lands located along coastlines, getting stuck and exposure to temperature and cold climate would cause fatalities. Again, avoidance to such regions is the solution. If not, do not venture into such regions alone; so that help would be available at hand.
- The quicksand is not so quick. The floating sand takes a few seconds to get disturbed before making a sudden downward settling of sands; the action is what actually "swallows" the victim. Hence, while walking on wet grounds, have a wooden stick, check for any unstable ground before setting foot.
- The sudden settling of sand particles to the bottom surface makes the remaining silty and clayey particles to become sticky, that traps the victims or body parts; extricating from which is too difficult. In such situations, addition of water makes the sticky material to loosen. Making slow, circulatory motions towards firm ground instead of panicky reactions is suggested for escaping from such danger.

References

Anonymous., 1984 Is Quicksand really deadly? Sci.Digest. p. 20.

http://en.wikipedia.org/wiki/Dry_quicksand.

http://en.wikipedia.org/wiki/Quicksand.

Khaldoun, A., Eiser, E., Wegdam, G.H. and Bonn, D., 2005 Rheology: Liquefaction of quicksand under stress. Nature. v.437. pp.635.

Leslie, W., 1990 Quicksand Makes a Good Story. Audubon. pp.126-132.

Lohse, D., Rauhé, R., Bergmann, R. and Devaraj van der Meer., 2004 Creating a dry variety of quicksand. Nature. v.432. pp.689-690.

Ramon, G. and Wessner, C., 1996 How Does Quicksand Work? Pop.Sci. p. 86.

www.Mythbusters.com.

❑❑❑

17

Glacial Lake Burst in the Lunana Area, Bhutan : A consequence of Global Warming

***O.N.BHARGAVA, **S.K.TANGRI, ***A.K.CHOUDHARY & ****Y.DORJI**

*103 Sector 7, Panchkula 134109, India.
E-mail: <onbhargava@gmail.com>
** Sector 33, Chandigarh, India. ***Hydroelectric Project Authority, Bhutan.
**** Department of Geology and Mining, Thimphu, Bhutan.

Owing to the dynamic interaction between atmosphere-lithosphere-hydrosphere-biosphere, any change in one of these components either natural or imparted by anthropogenic intervention impacts other speheres and that too on a distal place! This chapter explains how global warming affects remote places like highlands of Bhutan. Severe hot summers during the year 1994 led to excessive melting of glaciers in the higher Himalayas. The exceptional melting contributed enormous amount of melt water to the lake basins, which caused breach of the morainic dam and consequent enlargement of the outlet of the Luggye Tsho and the Tshopda Tsho, Bhutan. Floods followed the breach on October 7, 1994. These floods destabilized the slopes on either bank of the Pho Chhu near Tshopda Tsho outlet making them prone to landslides. The flash floods also eroded the left lateral moraine of the Raphstreng Tsho, which had disturbed the original angle of repose. The Raphstrcng area is also prone to ice avalanches, rock glaciers, surge of which can damage the rim and also cause spillover of water. Examination of geological, geomorphological, geotechnical, microclimatic and hydrographic information in and around the glacial lakes led to improved understanding of breach characteristics based on which mitigation measures were suggested and implemented in the years 1996-97. Since then the lakes have remained intact. From this case study and satisfactory performance of mitigation measures suggest that similar integrated studies have to be conducted for thorough understanding of hazard prone areas based on which mitigation measures have to be implemented for successful reduction of impacts.

INTRODUCTION

The Lunana area, situated in the northern district of Gasa, is one of the most extensively glaciated terrains of Bhutan (Fig. 1). In response to the general global warming, these glaciers like those of other parts of the Himalaya are in a state of steady retreat, often leaving behind a chain of lakes. Though the glacial lakes endow scenic beauty, they also pose environmental hazards, due to frequent bursting of their morainic dams and consequent flash floods. The last lake-burst occurred on 7th October 1994, leading to floods in the Pho Chhu River. Following floods, the Geological Survey of India in collaboration with the Royal Government of Bhutan, investigated the Lunana area to identify the causes and effects of the floods and suggest remedial measures. More than a decade had passed since the implementation of mitigation measures and there has been no such breach and flood despite continued Glacial Vetreat during to Global Warming. This chapter describes the characteristics of these glacial lakes, nature and results of the study and mitigation efforts.

GEOLOGY

Gansser (1983) and Bhargava (1995) described the geology of the Bhutan Himalaya. Excerpts from these studies and observations from present study are as hereunder.

The lower reaches of the study area dominantly expose migmatitic and biotite gneiss of the Thimphu Group, which are extensively intruded by 2-mica granite and the younger undeformed leucocratic tourmaline granite and pegmatite. The Thimphu Group is overlain in the upper reaches of the Table Mountain by biotite schist, quartzitic marble and calc-silicate rock assemblage representing the basal Tethyan metasediments (Gansser, 1983). This metasedimentary suite represents the southern edge of the wide Kempa Dzong basin of Tibet. This basin, in its basal part contains a number of tourmaline granite sills (Gansser, 1983). Pebbles of black shaly slate resembling the slates of the Lingshi Group (Mesozoic) occur near the Luggye Tsho indicating a probable

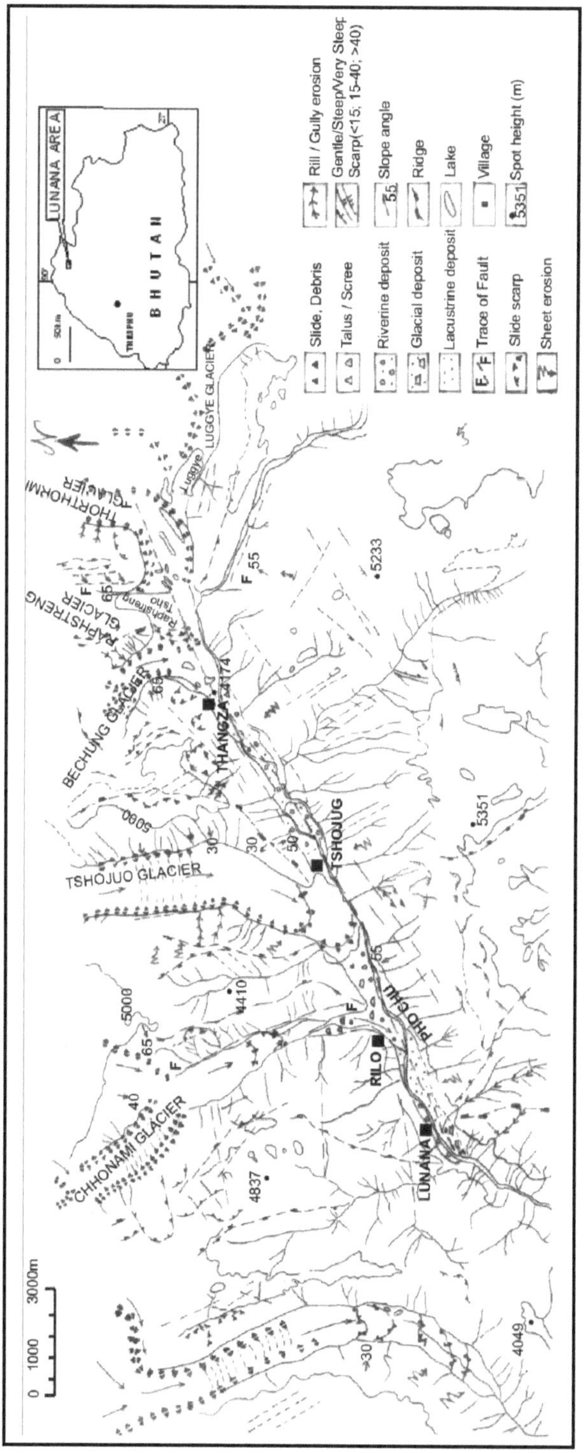

Figure 1. Photo- interpretation map of the Lunana area

existence of the Mesozoic sequence in the eastern extension of the Table Mountain Range. The Quaternary deposits of the Lunana area are of four categories, namely, glacial and glacio-fluvial, fluvio-glacial and fluvial, lacustrine and aeolian.

The Glacial and Glacio-fluvial deposits occur as six kame terraces along the Pho Chhu valley and as moraines. The fluvial sediments occur as flood plain deposits along the Pho Chhu valley. These are also preserved as channel deposits showing alternation of gravel and sand in the Tshoju flat and display upward fining and medium scale trough cross bedding. The fluvial sediments, since have undergone limited transportation, show moderately sorted subrounded to subangular clasts in a sandy matrix. Isolated patches of sand occur within the floods plain deposits. The lacustrine sediments are present at Thanza and Tshoju flats. The Tshoju deposit seems to have been formed due to ponding of the Pho Chhu waters by left latero-terminal moraine of the Tshoju glacier. Millimeter scale varves are present in the lacustrine sediments. Doubtful aeolian sand occurs as sandy layer within the lake deposits of the Tshoju area and also as local thin sand sheets and as small coppice dunes in the Tshoju flat.

Structurally the Lunana area constitutes the eastern continuation of the major Masa Kang antiform (Gansser, 1983). The regional structural pattern has been complicated by the presence of a series of NNE-SSW trending cross-folds. The granite and granite gneiss display six sets of closely spaced (10 cm to 2m apart) joints having diverse orientations as revealed by the plots of poles to the joints as recorded in the Raphstreng Tsho, the Luggye Tsho and the Tshopda Tsho areas. However, strike-wise, the master joints trend between NW-SE and NE-SW with moderate to steep dip towards NE and SE. The joints in majority of cases are oblique with respect to the orientation of the main foliation in the granite gneiss. Intersection joints causing block faulting have been noticed. Locally minor slips/faults occur along the joint planes. The earliest recognizable foliation plane in the granite gneiss and schist is the schistosity (S_1). The variation in the strike of the S_1 is due to the later cross folding having a NNE-SSW trending axial plane. The earliest F_1 folds are not preserved in the area. The F_2 folds have detached limbs. As revealed by their relationship with the S_1 plane, the F_2 folds are co-planer and also co-axial with the F_2 folds. The Pho Chhu between the Luggye Lake and Lhedi has a straight course. This course is neither parallel to any joint direction nor is there any fold axis located in the area. Possibly a fault exists in this area which controls straight course of the river. A N-S strike slip fault seems to have formed the Raphstreng Lake.

Neotectonic disturbances can destabilize morainic barriers and hence they were studied. Following observations indicate prevalence of neotectonic activity in the study area.

- Near Tshopda Tsho outlet, the bedding and fluvio-glacial terraces of the Thorthormi glacier have been tilted at an angle of 35°. The tilting is interpreted to be of recent tectonic event based on the fact that the tilted geomorphic surface, unaffected by the weathering, has retained the parallelism with the bedding and shows no signs of prevalent erosion since tilting event.
- One kilometer east of Shoju village, a synformal downwarp is present in the lacustrine sediments. This movement is obviously post-glacial, hence likely to be of Holocene age.

These neotectonic features, confined to the river valley, are discrete in nature. The fault along the Pho chhu, which controls its straight course, possibly was reactivated during the Holocene and caused the aforementioned features. The N-S fault seems to have exercised a parental control on the origin of the Raphstreng Lake. A strike-slip movement along this fault had created a 107m deep basin and also possibly severed the tip of the glacier, which remained confined to the north of Raphstreng Tsho.

GEOMORPHOLOGY

The altitude (above mean sea level) of the study area varies from 4200m at Thanza to more than 7000m at the Table Mountain. The topography of the area is still immature and is under active erosion. Most of the hills are structural-cum-denudational in nature. The landform is primarily controlled by lithology and structure. The study area can be broadly divided into four major geomorphic units namely, the glacial zone, the fluvio-glacial zone, the fluvial zone and the lacustrine zone.

The glaciers occupy higher altitude areas, while the adjacent low altitude areas show typical glacial geomorphic forms. All the glaciers of the study area exhibit a deflection towards northeasterly direction in their ablation zones that gradually shift to southeasterly direction in an arcuate pattern. The Tshoju glacier is the largest transverse glacier found in the Lunana valley. It possesses a significant large ablation area of nearly 10.5 km in length and occupies almost 52 km^2 area.

This glacier presently appears to be oscillating and is devoid of any lakes. The Chhunami glacier is the second largest tributary glacier found in the Lunana (Pho Chhu) valley that has suffered maximum recession amongst all the tributary glaciers. The Raphstreng glacier is the third largest tributary glacier in the Lunana region and its terminus directly debauches into the largest proglacial lake. The glacier flows towards south and it is 3.5km in length and 0.3km in width near its snout. It is almost perched on the southern slope of the Table Mountain encompassing the catchment area of 9.24km^2. The ablation zone is highly crevassed and serac faces are developed in all the directions rendering it unapproachable. The ice is free of supra-glacial morainic cover. Its snout is about 25m above the lake level. The glacier has receded by 250m between 1984 and 1995 having an average annual recession rate of 23m. Three distinct glaciarets coalesce together in the accumulation zone of the 'I'horthormi glacier, which is 5km long and 2km wide in its frontal part. It flows towards south. Avalanches largely replenish it. This glacier fans out towards southwest and its ablation zone is highly crevassed. Hummocky morphology has been imparted by serac development. This glacier is bounded by a thick pile of latero-terminal moraine of different stages in an arcuate pattern.

The Luggye north glacier occurs in the adjoining valley of the trunk glacier towards north. It is a small sickle-shaped glacier with a large ablation area. In its ablation zone, lateral morainic ridges on either side surround the glacier. The glacier bifurcates into two parts in the lower portion of its ablation zone. Luggye north glacier possesses the longest melt water channel that joins the Pho Chhu. The Bechung glacier is a small crescent shaped valley glacier. Its snout front exhibits significant recession. Lateral morainic ridges bound its ablation zone on either side and it is covered by supraglacial moraine. A small proglacial lake exists at two levels of the terminal moraine deposited by the receding glacier. Its melt water channel criss-crosses large morainic debris before its confluence with the Pho Chhu at 4174 m.

The fluvio-glacial zone, which exhibits the features of both fluvial and glacial origin, lies in regions located between 4200m and 3000m elevations. The glacial topography has been modified latter by the superimposed fluvial action, manifested by fluvio-glacial and kame terraces. The outwash plains are present at Tenchey-Thanza, Raphstreng and the Thorthormi complex. Beyond the terminus of the major glacier, the valleys are generally 'U' shaped. In this zone, numerous lakes of varying dimensions are present.

The fluvial zone characterized by the presence of river terraces, fans, fan-cut terraces and point bars is located in downstream portion of the major glaciated valleys and narrow linear valleys of the tributaries of the river Pho Chhu. The Pho Chhu with its source at the Luggye Tsho first flows due west for about 3.51km up to the south of Raphstreng Tsho and thereafter, it follows a southwesterly course for about 19km up to Shangsa. This river has a gradient of 1:26 in its upper reaches between the Luggye Tsho and Thanza village, 1:80 in the stretch between Thanza and Tshoju and further downstream up to Lhedi it steepens to 1:54. Six distinct levels of kame terraces corresponding to phases of glacial retreat are recognized in the Pho Chhu valley.

In the upper reaches, four glacial lakes are ensconced in the right side of the Pho Chhu valley and are situated at the snout of retreating glaciers. These lakes are: (from west to east) the Bechung, the Raphstreng, the Thorthormi and the Luggye Tsho. The Rapshstreng Lake is 107m deep and shows inverted profile i.e. rise in lakebed profile on either side. This type of erosion cannot be attributed to glacier; the glacier towards downstream could erode up to a depth of 107m (its bathymetry), but could not have moved up against gravity for about 100 m. This lake is rhomb-shaped towards its southern extremity. It is likely that the Raphstreng Lake was created as a pull-apart sag-pond due to a movement along a N-S strike-slip fault (*cf.* Reading, 1980). Except the Raphstreng Lake, other lakes seem to have been formed due to damming of the stream of receding glaciers.

The Bechung Tsho is the westernmost proglacial lake. It is situated about 1.5km northeast of Thanza village at an altitude of 4260m. It is a small shallow lake with maximum dimensions of 300m X 150m. The Bechung glacier that exists to its north feeds the lake. The lake has an outlet towards southeast. Almost N-S oriented Raphstreng Tsho is the largest glacial lake of the area at an elevation of 4350 m. It has a spatulate shape (rhomb in southern part) with a maximum length and width of 1.94 km and 1.13km respectively and a total surface area of 1.38 km^2. Bathymetric survey revealed that the lake has a gently undulating topography with a maximum depth of 107 m. A ridge divides the lake into two sub-basins. The lake floor is characterized by broad U' shaped valley with 60^0 to 65^0 slopes. The volume of lake water per meter depth is illustrated in Fig. 2. The volume of the water held in lake as calculated by east-west cross sections (Fig. 3) is 66.83 million m^3. The inlet point is in the extreme north, where Raphstreng glacier snout debouches directly into the lake. The lake water flows out through a 25m wide outlet through two subsidiary lakes. The prominent scarp located in the eastern segment of the lake might either represent lake terraces formed during successive rise in the lake level or due to the slumped morainic material.

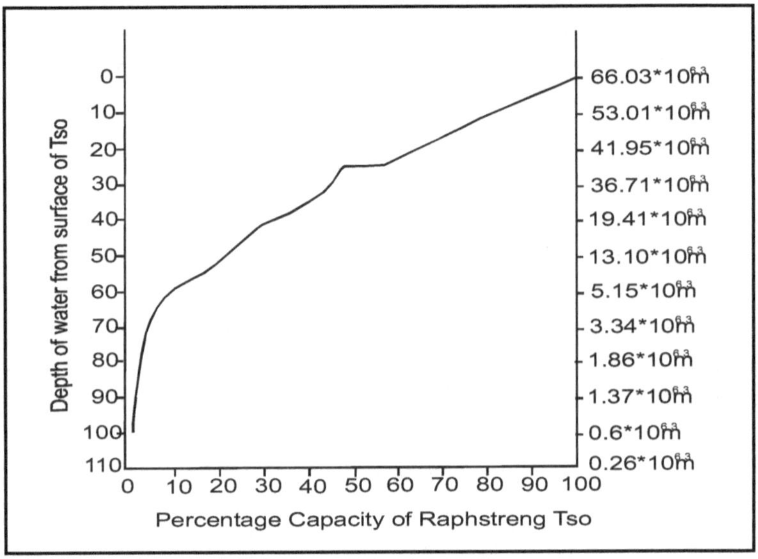

Figure 2. Graph showing depth-wise volume of water in Raphstreng Lake

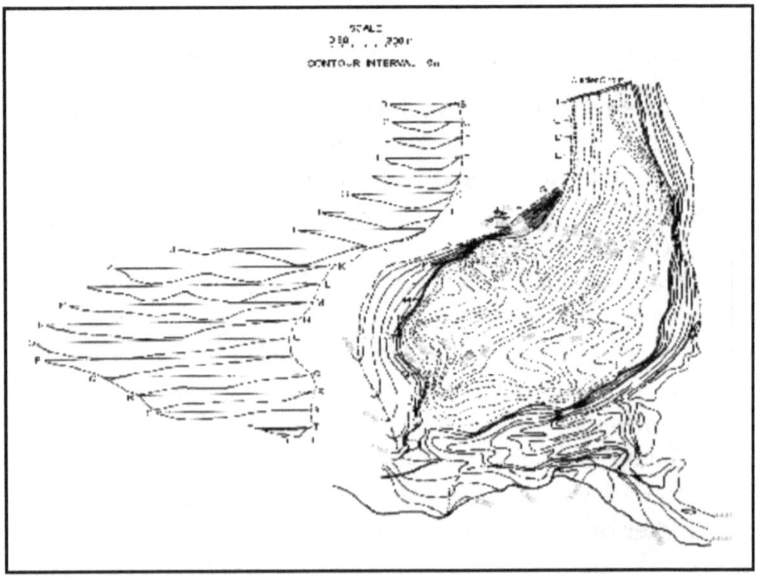

Figure 3. Bathymetric map and basin profiles

The first subsidiary lake has a surface area of 0.0023 km^2 and volume of water in it is of the order of 5750 m^3 with a maximum depth of 5.6m. The second subsidiary lake situated six meters below the first one at 4346 m has a surface area of 0.0066 km^2 and a maximum depth of 10.5m. The volume of the lake water is 2300 m^3. Water flows out from this lake through a small rivulet. The second subsidiary lake has a linear seepage zone of 50m in its southeastern part. The water discharge from the seepage zone is furnished in Table 1. The morainic barrier of the Raphstreng Tsho is approximately 4000m in length along its periphery. Its northern extremity is about 250m wide and abuts against the glacial ice wall. The lateral, medial and terminal moraines surround the barricade of reservoir (surface area of 1.38 km^2). The left- lateral moraine has a maximum height of 74m above the lake level in the southeastern sector, which gradually decreases to 1.5m near the lake outlet. In the SSE part, the left lateral moraine has a thickness between 295m and 410m. It is 178m in the SE part. The reduction in width is possibly due to erosion by the Pho Chhu during the 1994 flash floods. In this part, three subsidiary morainic ridges are identified. These ridges truncate against the right lateral moraine of the Thorthormi glacier suggesting that the Thorthormi glacier might have advanced after the advancement of Raphstreng glacier resulting in truncation of lateral moraines of the latter glacier.

A cluster of small lakes is situated at an altitude of about 4430m to the southeast of Raphstreng Tsho. The lakes are located at the terminal part of the fast decaying Thorthormi glacier. While few of these lakes are inter-connected through surficial water channels, others form isolated water bodies. The biggest lake located in the SE corner has a lens-shape and measures 300m and 60m along its longest and shortest axes respectively. The latero-terminal moraine of the Thorthormi glacier marks its southern border while the decaying glacial ice defines the remaining sides. Besides, several pools are situated amongst the ice floes of the decaying glacier. The density of these pools is higher in the north and NW parts of the Thorthormi glaciers, where the ablation rate of the glacier is higher. Falling ice floes from the marginal ice masses constantly bombard the lake. Upstream, the right lateral moraine of the Thorthormi glacier follows the hill-spur separating the Raphstreng and Thorthormi glaciers. Further downstream, it coalesces with that of the left lateral Raphstreng moraine, eventually truncating the latter. The left lateral moraine for major part of its length has a steep slope with maximum elevation difference of about 75m from the lake level. Except for SE segment, the remaining part of the moraine is well stabilized with thick bushy vegetation. The barren segment contains numerous landslides and exposes a small (3m x 2m) patch of dead-ice. The northern part of the Thorthormi left lateral moraine shows a big breach, possibly caused by the sudden advancement of the Luggye north glacier or gushing out of its melt water.

Table 1. Surface water discharge data of the Lunana Lakes

Date	Time in hrs.	Location	Discharge rate in M^3/Sec	Weather condition
07.7.95	1100	Raphstreng Tsho to last subsidiary lake	5.35	Cloudy weather, rains in the afternoon.
07.7.95	1700	Raphstreng *nala*	5.48	Cloudy weat her, rains in the afternoon.
15.7.95	1000	Raphstreng *nala*	5,71	Cloudy weather, rains in the afternoon.
15.7.95	1100	Raphstreng Tsho to last subsidiary lake	5.48	Cloudy weather, rains in the afternoon.
15.7.95	1400	Seepage zone of 2nd	1.77	Fair weather, sunny day .
15.7.95	1400	Raphstreng Tsho to last	6.70	Fair weather, sunny day.
15.7.95	1800	Raphstreng *nala*	6.87	Fair weather, sunny day.
19.7.95	1000	Raphstreng *nala*	5.17	Cloudy weather.
19.7.95	1100	Raphstreng *nala*	6.24	Sunny day.
19.7.95	1530	Luggye Tsho outlet	15.37	Cloud and rain day
24.7.95	1100	Luggye Tsho outlet	16.84	Sunny day, clear weather.
24.7.95	1400	Thorthormi outlet	7.28	Sunny day, clear weather.
24.7.95	1700	Raphstreng *nala*	6.79	Sunny day, clear wea ther.
29.7.95	0900	Raphstreng *nala*	5.38	Cloud day.
29.7.95	1000	Thorthormi outlet	6.80	Cloud day.
29.7.95	1130	Luggye Tsho outlet	15.37	Cloud day.
29.7.95	1530	Thorthormi Tsho outlet	6.97	Cloud day.
29.7.95	1700	Raphstreng *nala*	5.30	Cloud day
01.8. 95	0900	Raphstreng *nala*	5.69	Cloud day.
01.8.95	1100	Raphstreng Tsho to last	5.49	Fair weather.
01.8.95	1230	Seepage zone	1.82	Fair weather.
01.8.95	1500	Raphstreng nala	6.72	Fair weather.

The surface water outlet of the Thorthormi system of lakes is situated in the SW corner. The water first flows into two small shallow lakes, before flowing down into the Pho Chhu. *As* per the observation made earlier in 1983, the present outlet of the lake is new, the older being situated a little to the south of the present one. The water of the Thorthormi Lake is also muddy. The surface discharge of the lake is furnished in Table 1. No seepage through the morainic dam of the Thorthormi Lake is apparent.

Located below the majestic Jhumolhari peak, the rectangular shaped NW-SE oriented Luggye Tsho is the eastern most glacial lake of the east Lunana valley. It is situated at an altitude of 4480m southeast of the Thorthormi Tsho. It has a surface area of 0.9 sq.km. The Luggye Tsho is confined within the latero terminal moraines of the Luggye glacier, which defines the SE limit of the lake. The terminal morainic dam of this lake is comparatively much wider than the lateral moraines. The moraines have moderately steep slopes, which are well stabilized with bushy vegetation. The right lateral moraine has a maximum height of 102m above the lake level while the left lateral moraine rises to a maximum of 63m above the lake. Near the surface outlet of the lake, few dead-ice blocks are visible on lakeside of the left lateral moraine. The Luggye glacier snout extends into the lake. Frequently, ice blocks of various dimensions fall from the glacier into the lake. The lake water is muddy. The lake has reached its base level and its water flows out through a wide breached segment of the left lateral moraine. The surface discharge of the lake is given in Table 1. All through the length of the terminal and lateral moraines no seepage was recorded.

Tshopda Tsho is a crescent-shaped drained out lakebed, situated along the corner of the Pho Chhu valley, SSE of the main Thorthormi Lake. It is formed possibly due to the blockage of the narrow Pho Chhu valley by a landslide from the left lateral moraine of the Thorthormi glacier and the left bank of the Pho Chhu. As indicated by the level up to which recent sand has been deposited on the right bank of the Tshopda Tsho and also the water marks left on the rocky valley slope of the left bank of the Tshopda Tsho, the water level in this lake prior to bursting is deduced at 10m. Besides these glacier-fed lakes, there are two small clear water, surface run-off fed endmorainic lakes. The western lake is shallow 'L' shaped, situated in between the left lateral moraine of the Bechung glacier and the right lateral moraine of the Raphstreng glaciers. Eastern lake is comparatively smaller and oval shaped. It is located in between the left lateral moraine of the Raphstreng and the right lateral moraine of the Thorthormi glacier. The Thanza Tenchey and Tshoju dried lake beds forming lacustrine flats are situated at an elevation of 4174m. These deposits seem to have been formed due to ponding of the Pho Chhu water.

GEOTECHNICAL STUDIES

Five one cubic meter pits were excavated in the morainic material to carry out the geotechnical studies. Three were in the Raphstreng and one each in the Thorthormi and the Luggye Tsho. The excavated materials were subjected to grain size analysis, measurement of coefficients of uniformity and curvature, shear strength, permeability and unit weight.

The grain size and related data indicate that the morainic material is poorly graded, unsorted and contain lesser quantity of fine-grained sediments. The unit weight of the morainic material has been determined *in situ* for the calculation of overburden pressures. The unit weights in the three pits at the Raphstreng Tsho are 17.67 kN/m^3, 19.71 kN/m^3 and 19.66 kN/m^3 respectively. The values of unit weights of Thorthormi and Luggye areas are of the order of 24.66 kN/m^3 and 21.89 kN/m^3. The unit weight determinations are for the first one-meter depth from the surface up to which the exploratory pits could be excavated within the time frame and equipment available. The same ratio, if not more, is assumed for deeper regions.

Water percolation tests were carried out in the pits by falling head method. Permeability in the Raphstreng morainic deposits varies from 7.39 X 10^{-5} X 10cm/sec (7.39 Lugeon) to 8.36 X 10^{-5} cm/sec (8.36 Lugeon) indicating the material to be pervious. The permeability of the morainic material of Thorthormi and Luggye areas is of the order of 1.81 X 10^{-5} cm/sec (1.81 Lugeon) and 1.14 X 10^{-5} cm/sec (1.14 Lugeon) respectively, which indicate a highly impervious nature. It may be noted that excess of finer fractions in these two pits have reduced the permeability of the material. It is also corroborated by the unit weight of material in these pits. The permeability determinations done up to a depth of one meter in the pits, indicate that the material up to this depth is semi-pervious to impervious in nature. However, with the coarser fragments predominating at depth, generally, the morainic material can be considered as pervious for all practical purposes. The shear strength of the morainic material of the Raphstreng area has been determined by shear box test. The values of cohesion and friction angle are 0.02 kN/cm^2 and 370° respectively.

The discharge measurements (Table 2) were carried out between the period 7th July 1995 and 29th July 1995 in all the lakes. Measurements were conducted at a point where the water from the Raphstreng Tsho flows to the lower subsidiary lake and also near the Raphstreng Bridge. Discharge measurements in the Thorthormi and Luggye Tsho were carried out at the outlets. The melt water that feeds the lakes opens directly into the lakes by subglacial and englacial channels. A few small supraglacial channels in the ablation zone (Thorthormi) flowing along the length of the glacier also contribute melt-water discharge directly into the lake. It is observed that reduction of amount of flow of melt-water into the lake during night and substantial increase of flow during the day are due to diurnal variations of temperature and 'insulation'. Seasonal fluctuations in the melt-water flow of the temperate glaciers are well marked. Glaciers have the effect of storing winter precipitation and discharging it during the summer. During winter, most of the flow is due to basal sources while in the summer it is essentially due to surface flow (*cf.* Stenborg, 1995). As such, the amount of water discharge entering through subglacial and englacial channels that open directly into the lake remains almost constant throughout the year, whereas the contribution through supraglacial channels does not exist during the winter months.

Table 2. Discharge at different sites on 7th October, 1994

Gauge and discharge site	Distance from Luggye lake	Time in hrs. Sankosh river	Rise in level of
Lhedi	34km	0300	Not measured
Wangdue Phodrang	118km	0700	4.00m
Sankosh Dobani	188km	1040	2.40m
Kerabarl	242km	1420	1.84m

During mapping, two major seepage zones were observed. These are located in the frontal part of the lower subsidiary lake of the Raphstreng Tsho. The cumulative seepage from this zone is of the order of 1.8 cumecs. Water from the lower subsidiary lake appears to have followed the plane between the terminal moraines and found openings for seepages. These seepage zones are located between 4340m and 4344m. The morainic slopes, from where the water emerges, have an angle of 42^0 and support scanty vegetation. Due to seepage of water, the fine matrix of the morainic material is being washed away leaving voids between the coarser materials. The presence of the abandoned exit points at levels higher than the present ones, points to progressive downward migration of the seepage zones.

MICROCLIMATIC OBSERVATIONS

The microclimatic observatory was established at Thanza (4160m) to record maximum and minimum temperature, rainfall and dry and wet bulb temperature for computing the humidity. Minimum and maximum temperatures were recorded during the month of July. The minimum temperature ranged between 3.5^0C and 11^0C whereas the maximum temperature ranged between 14°C and 30°C. The lowest temperature of 3.5^0C was recorded on 24th July 1995 and the highest temperature of 30°C was recorded on 23rd July 1995. The area remains quite humid during July. The humidity ranged between 70% (on 8th July, 1995) and 100% (on 22nd July, 1995). During July, rains were quite frequent. Total rainfall recorded during July was 211mm. Maximum rainfall of 22.7mm in a single day was recorded on 30th June 1995.

These present observations together with microclimatic observations in the Tibetan plateau having past record of the glacial lake outbursts concluded that, in general, the lake outbursts occurred in years with higher average annual temperatures. Thus the microclimatic observations in the Lunana area assume significance. In an extended connotation, the trend of global warming could be implicated for a chain of events namely, glacial retreat-glacial lake formation-breach of lakes-flash flood hazard in the downstream regions.

STABILITY OF THE MORAINIC BARRIER

The stability of the morainic barrier, to a larger extent, depends on the horizontal thrust exerted by the hydrostatic pressure of impounded water column against the face head of the morainic barrier. The total hydrostatic pressure of the water column in the reservoir and the weight of the barrier section at the weakest point, where the width of the morainic dam has been reduced to 178m, were calculated as given hereunder. The morainic dam, at this point (Fig. 4) has a base width of 300m and the difference between the surface water level of the lake (4348 m) and the base level of the river (4322m) is 26m.

Volume of the morainic dam	=	(300 + 178)/2 x 26 m	= 6214 m^3/m length
Weight of the morainic dam	=	(300 + 178)/2 x 26 x 19.71	= 122477.94 kN
Hydrostatic pressure exerted in this part of the morainic dam	=	68132.96 - 51799.63 = 16333.33 kN	
Weight of the barrier/ hydrostatic pressure	=	7.49	

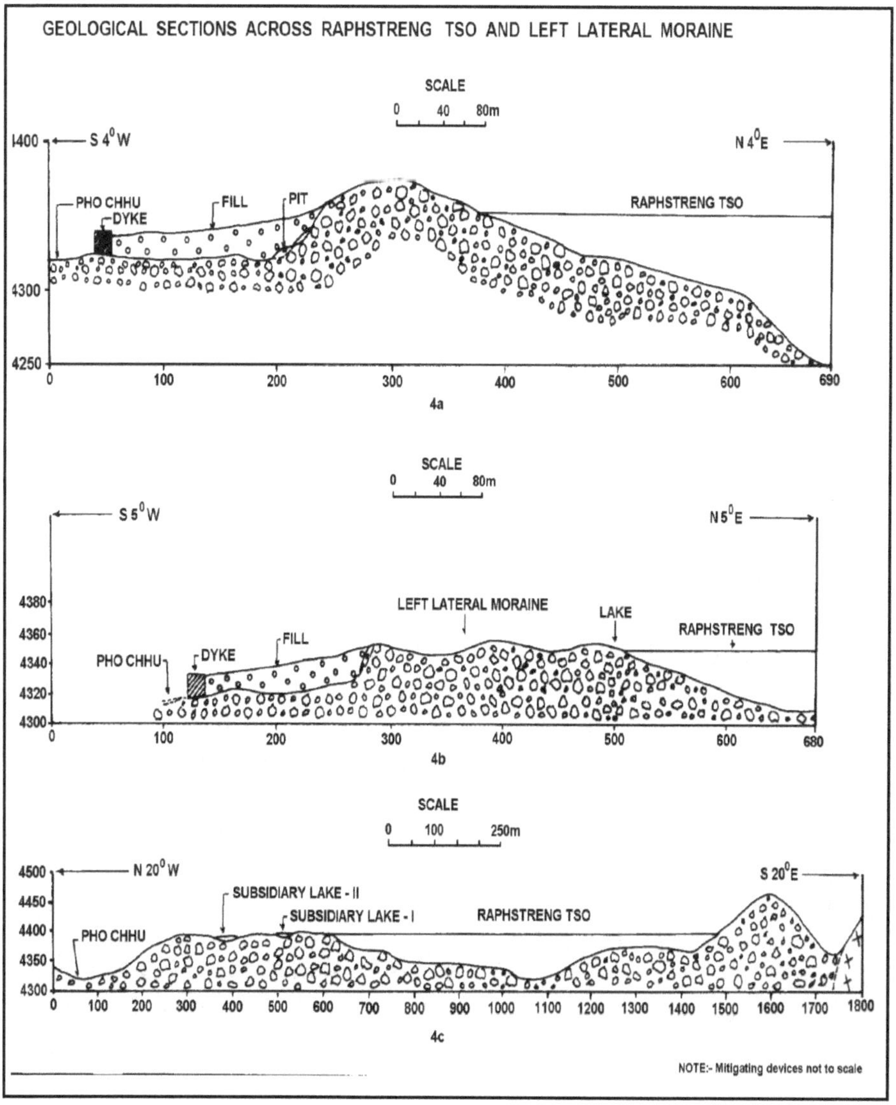

Figure 4. Geological Section of the Raphstreng Tsho and Left Lateral moraine

This ratio of 7.49 as worked out is based on the existing morainic dam profile. The original profile of this morainic dam before erosion due to last floods and formation of the embayment is likely to have a width of 400m against the existing 178m. The erosion by the Pho Chhu has reduced the ratio from 12.86 to 7.49 i.e., a reduction of 45% of original width of morainic dam prior to flooding. This ratio has been worked out without providing for 'dead ice mounds', in the morainic material. The dead ice noted in the year 1984 (Sharma et al. 1986) has since melted. With the above presumption, the safety factor (ratio) gets further reduced making the morainic 'barricade' still more vulnerable to failure on the hydraulic considerations or in the event of occurrences of earthquakes, impact of avalanche and cloud bursts, etc. The failure may rupture the arcuate morainic barrier of the Raphstreng Lake and release water volume of the order of 30.71 million cubic meters thereby causing havoc further downstream. Needless to mention that in due course of time even natural process of erosion shall breach the barrier.

LAKE BURST AND FLOODING IN THE LUNANA AREA

The bursting of glacial lakes in the Lunana area, like other glaciated terrains the world over had a devastating effect on the ecosystem. Though the lakes must have burst in the remote past also as is evident from the Tshoju lake sediment, no records of floods are available prior to 1950. A major flash flood lasted for about 24 hours and damaged a part of the Punakha Dzong in the year 1950 (Gansser, 1970). As a follow-up, Gansser (1978) made suggestion regarding investigation of endangered lakes. The origin of these flash floods was traced to sudden draining (i.e. bursting) of the Tarina Lake, in the western Lunana area, due to breaching of the terminal moraine, which had dammed the lake. The breaching was caused by the impact of an ice-avalanche, which day lighted from an overhanging glacier. It was followed by another major flood during the year 1960. This event was more disastrous than the 1950 floods. Its origin was attributed to bursting of a lake in the eastern Lunana area. The floods lasted for five days. The floods of the Sankosh River in 1968 caused extensive damage to the Wangdue Phodrang Bridge. Loss of human life was also reported. On 7th October 1994, catastrophic flash floods played havoc throughout the Pho Chhu valley. The toe cutting by floodwater had initiated landslides between Luggye and Lhedi, over a distance of 35km, prominent being about 70m downstream of Tshopda Tsho and creep movement along the thalweg of the river system. Sizeable fauna and flora and many human lives perished in the floodwater. Along the western bank of the Pho Chhu, dead- ice was exposed in the left-lateral moraine damming the Thorthormi Tsho. Left medial and lateral rims of the Raphstreng, made up of morainic material were eroded, reducing its width to 178m (Fig. 4) as compared to 410 m towards its upstream. Further downstream, the floodwater had eroded part of pasturelands located between Thorthormi, Raphstreng and the Bechung lakes and agricultural and pasturelands between Thanza and Lhedi. It also damaged a part of the Dzongchung in Punakha. During a preliminary appraisal of the area, Anon (1994) attributed the cause of 1994 floods to the bursting of the Luggye Lake. The total discharge as a result of this flood, recorded at various sites downstream of Luggye is furnished in Table 2.

An hourly monitoring of flood discharge was made at Kerabari. At its peak, 1654 cumecs of water was recorded at 14.20hrs. on 7th October 1994. It took 18 hours for the floods to subside, during which period, a cumulative discharge of 23 million cubic meters was recorded at the discharge site. The damage due to floods was confined to the eastern Lunana area wherein the morainic dams of the Bechung, Raphstreng and the Thorthormi Tsho were found intact. The morainic dam of the Tshopda Tsho, formed by the left latero-terminal morainic ridge of the Thorthormi glacier was found breached and lake was completely drained out. There were also evidences of deposition of the recent sand and of wave action in the Luggye Lake, whose rim was also found breached and water level brought to base level. A breach in the Luggye rim was observed even in the year 1984. The pre-1984 breach was obviously enlarged during 1994. The widening of breach could be due to melting of the dead-ice (present even now) due to excessive hot summers, which might have led to piping action, removal of finer fraction of sediments and consequent weakening of the morainic material which finally yielded to hydrostatic pressure. As large volume of water was released from the Luggye, its erosive power breached the downstream of Tshopda Tsho dam also.

POTENTIAL HAZARD ZONES

The destabilized areas, especially close to the Tshopda Tsho outlet, have become prone to landslides. The landslide has partially blocked the Pho Chhu course. Damming of the Pho Chhu valley due to slide would lead to formation of a lake, saturation of morainic and landslide materials with water, both of which shall loose shear strength, causing slope failure. Though presently the Luggye Tsho has acquired its base level, yet as probably happened between 1984 and 1994, the breach can be healed and with accumulation of large volume of water, a tragedy akin to that of 1994 can recur.

The Pho Chhu valley bed, at Tshopda Tsho outlet, is situated at a level higher than that of the Thorthormi Tsho. The Pho Chhu River, in the Tshopda Tsho reservoir will exert hydrostatic pressure. Also the water from higher level may seep through the morainic material, especially along its contact with the dead-ice into the Thorthormi Tsho due to increased temperature and consequent greater melting; the resultant increased hydrostatic pressure may endanger stability of its morainic barrier. With accelerated rate of melting of glacier during excessive hot summer years, several pools of water of the Thorthormi complex can merge to form a large lake with effects enumerated above.

Left lateral moraine of Raphstreng Tsho has a tapering width. Its 410 m width at upstream, possibly due to erosion during 1994 floods, has been reduced to 178 m. The toe erosion of the morainic material has disturbed the original angle of repose. The area, besides further erosion, is also prone to ice-avalanches and rock glaciers, surge of which can damage the rim and also lead to spill over of large quantities of water. The seepage in the lower subsidiary lake of the Raphstreng Tsho has been recorded at a low head of 7m with a creep length of 10m to 20m. The seepage is also relatable to the surface outflow of the Raphstreng Tsho. The seepage as determined during the present investigation is of the order of 1.8 cumecs. The seepage in turn takes away the fines and scours the toe of the morainic barrier that shall ultimately destabilize the slope.

MITIGATION MEASURES

Following short term and long term mitigating measures were suggested to ameliorate the fragile ecosystem of the Lunana area (Fig.5).

Short Term Measures

- The free flow of Pho Chhu water in Luggye-Tshopda Tsho sector should be ensured by unloading of crown of the active landslide situated above 70m downstream of the Tshopda Tsho outlet.
- Gabions-toe training walls to be constructed on either bank of the Pho Chhu (Fig. 5).

- The outlet of the Thorthormi Lake may be lowered by 10m to release the water and reduce the hydrostatic pressure within the safe limits. The lowering of water level should be done in stages to avoid any damage to downstream regions.
- Restoration of the original toe section of morainic dam washed away by the last floods with the formation of an embayment should be achieved by filling that portion by morainic material/ large sized boulders encased in wire mesh. In order to prevent further erosion and to chanalize the waterway of the Pho Chhu, a set of gabion/dykes as shown in Fig. 5 may be constructed. Abundant stones are locally available for the construction of such dykes. Wire mesh of suitable size shall have to be transported.
- Lowering of the water level to 20m (Fig.5) shall result in fall of the volume of water by 38% and hydrostatic pressure from 68,152 kN to 51,799 kN. The lowered level shall yield a 1:10 slope to the morainic dam. The reduced hydrostatic pressure shall provide stability to the morainic dam and also minimize the chance of spillover or overtopping due to surge, should ice avalanche/rock glacier daylight from the main glacier.

Figure 5. Three-dimensional sketch of the Raphstreng Tsho showing mitigating measures

- Siphoning and auxiliary spillway cutting may be carried out simultaneously to reduce the water level. Any suitable pipe (e.g. polyvinyl), which can stand low temperature or freezing of water inside, may be deployed to drain the water. The draining should be kept at a rate of 10 cumecs. Simultaneously, the spillway may be progressively cut. The spillway may have a width of 40m with 1.5 h/v side slopes.

Long Term Measures

During catastrophic events like earthquakes, the short-term measures may fail without any warning. In such a case, the downstream habitations are likely to be ravaged by the floodwaters and associated landslide, ice avalanche, etc. Hence, following long-term measures are suggested:

- To meet such an exigency, it is utmost important to strengthen/ protect buildings, villages and important fields by conventional and also innovative civil engineering techniques. Construction of check dams, river training dykes and other energy dissipation devices may be established at Thanza, Tenchey, Tshoju and Lhedi to chanalize the river and to add stability to the valley walls. In addition, some of the habitats that are located very close to riverbed may be suitability relocated away to guard against flash floods. The Punakha Dzongchung, which for historical reasons cannot be relocated, requires special protective measures.
- Plantation of fast as well as slow growing grass/shrubs/trees, having soil holding capacity may be undertaken to stabilize the slopes.
- Regular monitoring of the area to record increase/decrease in lake levels, formation of artificial dams in Pho Chhu and disstress condition of valley walls should be ensured during rainy season and during the months of April to October when glaciers melt.
- The absence of seismic records of the area makes it difficult to assess the effect of any seismic event to the morainic dam. Neotectonic activity recorded in the area has been attributed to the possible activation of Pho Chhu fault and a strike-slip fault along the Raphstreng. In view of these observations, and fairly high degree of seismicity being associated with N-S transverse features in the adjoining Nepal-Sikkim Himalaya, establishment

of seismic stations is considered necessary. The data would be helpful in determining the seismic status of the neotectonic features in Lunana lake area.

- Though these recommendations are specific to the east Lunana area, similar measures can be implemented in west Lunana area, as well as in other glaciated terrains. In the Tarina area, where the lake has acquired the base level, the free flow of the water should be ensured so that no damming recurs.
- Since a relation between the lake-bursts and hot summers has been established, it is advisable to set-up a meteorological observatory in the Lunana area to record micro-climatic variations.

These measures suggested in 1995 were implemented during 1996 and 1997. Since then, all the lakes have remained intact.

References

Anonymous., 1994 Preliminary Report on the investigation of Glacial Lakes at Pho Chhu source and the assessment of flood affected areas in Lunana. Royal Government of Bhutan (Unpublished Report).

Bhargava, O.N., 1995 The Bhutan Himalaya: A Geological Account. Geol.Surv.India. v. 39. 245p.

Gansser, A., 1970 Lunana, the Peaks, Glaciers and Lakes of Northern Bhutan - The Mountain World, Swiss Foundation of Alps. Res. Zurich. pp.117-131.

Gansser, A., 1978 Proposals for an investigation of dangerous lakes in North Bhutan. Royal Government of Bhutan (Unpublished Report).

Gansser, A., 1983 Geology of the Bhutan Himalaya. Birkhauser Verlag, Basel. 181p.

Reading, H.G., 1980 Characteristics and recognition of strike-slip fault systems. *In:* Balance, P.F. and Reading, H.G. (Eds.), Sedimentation in oblique-slip Mobile Zone. Inter.Assoc. Sediment.Spl.Pub. v.4. pp.7-26.

Sharma, A.R., Ghosh, D.K. and Norbu, P., 1986 Report on Lunana Lake Expedition. Geol.Surv.India.Bhutan Unit. (Unpublished Report).

Stenborg, T., 1985 Glacial Morphology, John Wiley & Company, London. 2nd Edition. 362p.

18

An Attitudinal Approach in Management of Diasaster Mitigation and Risk Reduction

R. SUBRAMANIYA BHARATHY
Periyar Institute of Management Studies, Periyar University, Salem - 636 011
E-mail: <rsbharathi78@yahoo.co.in>

"Those who would take over the earth and shape it to their will Never, I notice, succeed.The earth is like a vessel so sacred that at the mere approach of the profane it is marred and when they reach out their fingers it is gone" - Chinese Philosopher Lao Tzu Tao Te Ching. True to his verses, geohazards are phenomena regulated by the law of nature and it is beyond the human intellect to understand thoroughly and capability to conquer. This is the lesson learnt by us from so many incidences that geohazards cause greatest havoc wherein natural processes are going on and are inhabited by humans. Greed for natural resources and population pressures on the earth's resources are the causes of many a catastrophic event. Examples in this regard are galore including, damning river courses and location of urban centers along rivers and coasts – flooding, cyclonic storms, tsunami hazards; alteration and deforestation hill slopes for agricultural and construction activities – landslides; unscrupulous utilization of ground water resources – drought, etc. This chapter is an attempt to draw the attention of human community as a whole to turn towards learning to live with the nature without alterning the natural processes to get rid of impacts of geohazards as a best ever mitigation measure. In other words, this

chapter attempts to emphasize the need to change our attitude to escape from natural hazards instead of attempting to tinker around natural processes only to get battered at the end.

INTRODUCTION

Oxford English dictionary states that the word 'disaster' is derived from the 16th century French word 'disastre', which is a combination of two terms namely, 'Des' meaning bad or evil and 'astre' means star. Thus "desastre" signifies a 'bad star' or 'evil star'. Disaster, therefore, implies loss or damage occurring due to some unfavorable star. However, in terms of modern knowledge, the word 'disaster' denotes any odd event, whether natural or man made, which can bring about sudden and great miseries to humanity in the form of loss of life or/ and property.

India has been traditionally vulnerable to natural disasters on account of its unique geographic and climatic conditions. Floods, droughts, cyclones, earthquakes and landslides are recurrent phenomena. These geological processes have the potential to cause great harm. According to an estimate, about 58% of the Indian landmass is prone to earthquakes of various intensities; over 40 million hectares of land is prone to floods; about 8% of the total area is prone to cyclones and 68% of the area is susceptible to drought. The loss in terms of private, community and public assets has been enormous. In this scenario, it is imperative that India as a country should learn to live with these hazards and make all necessary measures to minimize the ill effects of natural geological processes. However, we as a developing country with enormous population, always search newer resources to utilize and in the process value resources more than human lives. That is how recurrent geohazards and resultant lives lost are not being tackeled properly. There is another side of this paradigm. With the opening-up of economy and growing population, demands are on the rise to meet the needs of people, which, end up seeking newer resources and greedy utilization of known reserves. These two endeavors namely, economic prosperity and needs of growing population invariably result in extensive colonization of newer land resources in terms of urban centers, infrastructure developmental projects, commercial and recreational utilities. In this process, traditionally vulnerable and environmentally sensitive regions are being put to use - to which, the nature retaliates through many catastrophic events such as flooding, landslide, storm surge, melting of glaciers and resultant flash floods, etc. Regardless of life and property loss arising out of such catastrophes, the process of colonization goes on unabated.

While utilizing the natural resources for meeting the needs of growing population and for working towards economic prosperity of our country could not be disputed, it is unwise to have second thought on minimizing, if not avoiding economic and life loss arising out of natural calamities - both of which work against the very nature of fulfillment of needs and economic requirements. A balanced approach is the key word that is going to be emphasized in this chapter, which calls for a change in attitude of our society as a whole in managing geohazards, and maintaining natural equilibrium of geological processes.

IMPACT OF DISASTERS

Disasters are not totally discrete events. Their possibility of occurrence, time, place and severity can be reasonably and in some cases accurately predicted by the existing technological and scientific advancements. It has been established that there is a pattern in the occurrences of many geohazards. This awareness demands the study of disaster management in an orderly approach. The degree to which a socio-economic system or physical assets are either susceptible or resilient to the impact of natural hazards is determined by a combination of several factors, including

- Awareness of hazards,
- The condition of human settlements and infrastructure,
- Preparedness of public and administration,
- The wealth of a given society and organized abilities in all fields of risk management.

All these factors either directly or indirectly result in or connected to a single important factor, namely human - the root cause for transformation of a natural process into a disaster!

Lack of awareness among the public and decision-makers about factors and human activities that contribute to environmental degradation and vulnerability to disasters is aggravating the intensity of the impact of hazards. For example, river floods are aggravated or even caused by deforestation, which causes erosion in headlands to clog rivers at downstreams leading to overflow even during normal rains. The accelerated and often uncontrolled, growth of cities have contributed to the ecological transformation of their immediate surroundings such as pressure on scarce land, exploitation of scarce groundwater and forest resources, etc. In addition, lack of appropriate drainage systems and extensive sealing of percolation sites through the use of concrete and asphalt on land surface increases the volume and rate of rainfall runoff, making many cities more vulnerable to flash floods. Recent catastrophic earthquakes and tsunami event highlight other key deficiencies in the approach to disaster risk reduction, such as poor understanding by common public, lacksoidal approach of administrative machinery to violations on environmental protection acts and violations of building norms as well as the tendency of some builders, to use the cheapest designs and construction materials to increase short term economic returns on their investment. Similarly, notwithstanding the coastal zone regulation act, too many construction structures are located within the no-no zone from coastline all along the fringes of Indian landmass, which inturn, single handedly contributed towards death and property loss during the recent tsunami. The natural defences along coastlines created by the natural processes themselves were also destructed by the human community, such as leveling beachridges, clear-cutting of coastal mangrove forests, etc. which in turn have aggravated the vulnerability levels. Examples can be cited like this endlessly. For the sake of simplicity, it can be summarized as "the intervention of anthropogenic activities over natural processes and resultant transformation into disasters generates repercussions in terms of economic, political, social, cultural, ecological, sustenance, physical and moral values and institutional instability". Hence, a concerted effort to manage the ill effects without any of these repercussions is the need of the hour.

DISASTER MANAGEMENT STRATEGIES

Disaster risk management strategies have global ramifications and hence policymakers have started restructuring institutional setup of governments and the roles and responsibilities of government machinery ranging from coordinated efforts of nation states percolating down through individual countries, state, district and village levels. The restructuring process and coordination efforts include a wide array of activities such as

- Institutional reforms,
- Improved analytical and methodological capabilities,

- Education and creation of awareness,
- Financial planning and political commitment and social involvement.

Disaster reduction is aimed at motivating societies at risk to become engaged in the conscious management of contributive factors to reduce risk and vulnerability. As the traditional response in terms of providing relief and rehabilitation and establishment of defense mechanisms against the impact of natural hazards is not offering immunity, an ongoing process that does not focus only on singular disaster event but on the contributive factors that transform natural geological processs into a disaster. Based on this principle, four objectives have been formulated by the United Nations during the International Decade for Natural Disaster Reduction (IDNDR, 1990-99). These objectives listed herein set the stage for the course of action for governments, regional bodies and civil society organizations.

Obtaining the commitment from public authorities

This objective is addressed through an increased inter-sectoral coordination, development of risk management strategies and allocation of resources including creation of new funding mechanisms. According to this objective, disaster reduction is to be dealt as a separate policy issue and with a coordinated effort of various governmental departments such as public works, rural development, health, agriculture, food security and environment, etc.

Increasing public awareness and public participation on vulnerability reduction

This objective involves programs related to formal and non- formal education. It is addressed through public information, education and multi-disciplinary professional training.

Stimulating inter-disciplinary and inter-sectoral partnerships

It is achieved through the expansion of risk reduction networking amongst governments at national and local levels, greater involvement of the legislators, private sector, academic institutions, NGOs and community based organizations. It calls for strong coordination mechanisms such as appropriate institutional structures for disaster management, preparedness, emergency response and early warning, as well as the incorporation of disaster reduction concerns in national/

state planning processes. Efforts to link natural resource management with disaster reduction are also encouraged.

Fostering better understanding and knowledge of the causes of disasters

It is achieved through the transfer and exchange of experience and establishment of mechanisms for greater access to relevant data and information to all the stakeholders. The issues to be addressed in this context are the assessment and analysis of socio-economic impacts of disasters, disaster databases, coping strategies of different social groups, early warning processes, as well as the promotion of scientific research, valuing of indigenous knowledge and the development and transfer of knowledge and technologies.

DISASTER PREVENTION PLANNING

It is all about preemptive planning rather than piece-meal and post event efforts. Disaster prevention is thus, can be accomplished through following steps.

- Introduction of comprehensive process of vulnerability and risk assessment.
- Building a robust and sound information database containing data on land use, demography, infrastructure developed at the national, state and local levels along with current information on climate, weather and man-made structures is crucial in planning, warning and assessment of disasters. Resource inventories of governmental and non-governmental systems including personnel and equipment help in efficient mobilization are to be established. In addition, response measures have to be optimized.
- Creating state-of-the-art infrastructure and dynamic assessment of vulnerability levels of various terrains are essential in tackling relief and rescue operations. The entire disaster mitigation game plan must necessarily be anchored to frontline research and development in a wholistic mode. State-of-the art technologies available worldwide need to be made available in India for upgradation of the disaster management system. In addition, specialized research activities should be encouraged for devising technologies to improved forecasting, information dissemination and rapid mobilization of resources for relief operations.

- Establishing linkages between knowledge-based institutions with decision centres, policy makers and common public is necessary for the benefit of common public as well as robust economic growth of our country. Any measure capable of reduction of costs of relief and rescue operations, economic loss arising out of disasters, loss of agricultural and capital good production would greatly help save the precious monetary resources for the betterment of our nation's economy as a whole.

TECHNIQUES FOR EFFECTIVE DISASTER MANAGEMENT

Lessons learnt from disaster management schemes of other countries as well as our own experiences have sharpened the methodologies and procedural systematics involved in this task. Following are few of those techniques that may be put to use.

- Integrating disaster management into development planning.
- Multi-hazard approach to disasters.
- Leveraging existing Government machinery.
- Effective inter-agency cooperation and coordination.
- Capacity building.
- Autonomy and equity.
- Legal sanction.
- Accommodating aspirations of people.
- Tuning the measures according to local conditions.
- Financial sustainability.
- Cost sharing and cost recovery.
- Developing, sharing and disseminating knowledge.

Despite these strategized actions, well-meaning intentions of international community and concerned governmental machinaries, death-tolls of recurrent geohazards elsewhere, including developed countries such as U.S.A. (recent cyclonic devastation), developing countries such as India (flooding in Bihar), underdeveloped countries (drought in many of the African countries) raise a basic question – how effective all these measures are? Though the government has come with a comprehensive disaster management policy and system, it is up to the people to make it a successful one. The poor people are the most vulnerable to hazards/ disasters as they posses fewer resources and have very poor capacity to prevent or cope with the impacts. If we analyze the reasons, it is not because of the defective policy or the system but because of the "attitudinal factor" which played a major role in making the disaster management a failure if not less-successful one.

ATTITUDINAL APPROACH IN DISASTER MANAGEMENT

It involves developing appropriate attitudes in the individual. It is considered as the "most difficult task" that requires enourmous resources, persistence and above all, voluntary involvement from the stakeholders concerned. Not only the awareness, but also a mental state and a habitual procedure are to be inculcated for effective disaster reduction. This realization has dawned on the policy makers that introducing disaster management strategies or system is only a temporary measure without active involvement of ALL the stakeholders.

Biggest obstacle in this endeavor is that people do not want to change. Human being as an individual entity likes doing things the way s/he has learnt/likes to do as each individual is accustomed to specific habits, thinking patterns and the way of life. Some people are aware that their habits, thinking patterns and actions that are vulnerable to disasters, but still they refuse to change. They do not want to do this because it requires commitment. It may entail more effort or discipline or even letting go of something for which they have worked hard.

Recognizing that awareness about vulnerabilities is a *sine qua non* for inducing a mindset of disaster prevention, mitigation and preparedness, the Government of India has initiated a nation-wide awareness generation campaign as part of its overall disaster risk management strategy. Information on geohazards and their mitigation measures were included in school syllabus. Parliamentarians are being trained/informed on the nature and consequences of geohazards. Furthermore, in order to devise an effective and holistic campaign, a steering committee for mass media campaign has been constituted at the national level with due representation of experts from diverse streams of communication. The Committee has formulated a campaign strategy aimed at changing peoples' perception of natural hazards and has consulted the agencies and experts associated with advertising and media to instill a culture of safety against natural hazards. Apart from the use of print and electronic media, it is proposed to utilize places with high public visibility viz. hospitals, schools, railway stations bus terminals, airports, post offices, commercial complexes and municipality offices etc. to make people aware of their vulnerabilities and promote creation of a safe living environment. A novel method being tried is the use of frequently used stationery viz., postal letters, bank stationery, railway tickets, airline boarding cards and tickets etc. for disseminating the message of disaster risk reduction. Slogans and messages for this purpose have already been developed and have been communicated to concerned Ministries/agencies for printing and dissemination. The mass media campaign will help build the knowledge, attitude and skills of the people in vulnerability reduction and sustainable disaster risk management measures in due course of time. Even then, the real change will occur only when the people commit themselves to change. Following are few strategies people can adopt so as to develop a right attitude, which will definitely help save not only their personal lives and property but also of their surroundings.

- Avoiding the hazardous terrains; for example, by building structures and lifelines out of harm's way or by evacuating a threatened area.
- Building structures to withstand the effects of the hazard.
- Avoiding the environmental degradation and a sense of social responsibility.
- Improving the socio-economic conditions.
- Respecting the legislative instruments and the disaster management policy.
- Ignoring, manipulating or corrupting the codes should be avoided.
- Make use of the knowledge gained through the awareness and educational programs and to disseminate the same to the needy.

COURSE OF ACTION TO INCULCATE ATTITUDINAL APPROACH

Although resistance/reluctance to change is the nature of human mind, a little bit of shoving and mass-based programs, conducted, articulated and or followed by persons prominence has been found to be effective in inculcating habits among common people. In this regard, the policy makers can play a very vital role for fulfillment of the objectives of the Disaster management. The following areas should be considered as key elements for developing right attitude among the people.

- Drafting appropriate legislative system through required Instruments is an important task the government should accomplish to fulfill the objectives of the disaster management program.
- Capacity building and strengthening of institutional arrangements at all levels to address risk reduction as an ongoing function, including enactment of legislations for disaster reduction, restrictions on landuse, strict enforcement of building codes and reinforcement of environmental policies.
- Capacity building at state level needs to include the development of an integrated disaster risk management plan that covers areas of risk assessment, early warning systems, training programs, as well as emergency response management, recovery resources and strengthening of society based organizations.
- Development of public educational programs and campaigns on the relationships between sustainable development, natural hazards, vulnerabilities and disaster to enhance the effectiveness of disaster reduction measures. The process starts in schools with educational programs including curricula revision, training to teachers and development of resource centers. It needs to expand to all levels of society by training efforts, with special emphasis on professionals and community based leaders and organizations.
- Creating and implementing comprehensive urban development strategies and land use plans, which help mitigate damages caused by hazards. As location is the key factor determining the level of risk associated with a hazard, land- use plans and mapping should be used as tools to identify the most suitable usage for vulnerable areas (e.g., location of buildings, roads, power plants, storage of fuels). Local governments need to play an increasing role with regard to factors such as building standards, specific measures to be undertaken prior to or during or after construction projects.
- To improve resistance to the effects of natural hazards, better building practices are essential. Until there are building codes with modern technical standards and the force of law, they will not be taken seriously by the construction industry.

- Enforcement of existing codes effectively is the key to success but remains to be a point of sore until now. It implies that legally mandating codes are not enough; but the fruits of which would riphen only by strict adherence.

References

Arya, S., Karanth, A. and Agarwal, A., 2005 Hazards, disasters and your community. A primer for parliamentarians. Government of India. 62p.

Hamilton, R., 1997 Report on Early warning capabilities for geological hazards. IDNDR Scientific and Technical Committee Conference, Geneva.

Sundar, I. and Sezhian, T., 2007 Disaster Management. Sarup & sons, New Delhi.

❑❑❑

Glossary on Geohazards

***R.SURESH & **V. THIRUKUMARAN**
*Department of Geology, Periyar University, Salem - 636 011.
**Department of Geology, Government Arts College for Men, Salem - 636 007.
E-mail: <calcite15@yahoo.com>

In this section, we have presented a list of terms related to geohazards and attempted to provide explanation to them. These terms are collected from various sources not limited to but inclusive of those listed in references. While we thank the sources for their permission to utilize this valuable information, we undertake that the correctness of the terminologies and their meaning is the sole responsibility of the authors alone and not necessarily that of sources.

Accelerograph : Name of an instrument that records the vibrations of earth surface.

Accretionary wedge : Pile of sediments accumulated in subduction zones located between collisional margin of oceanic and continental plates.

Acid rain : Meteoric precipitation with dissolved components drawn from atmospheric pollutants. By dissolving gaseous materials present in atmosphere supplied by automobile, domestic, industrial and other sources, the rainwater becomes weak carbonic acid, nitric acid and sulphuric acid and affects the natural environments.

Active volcano : A volcano, which is currently erupting or has erupted in recorded history.

Active fault : Faults are considered to be active if they have moved during historic times or showing recent evidences of movement along fault plane and or generating seismicity.

Active Beach : Dynamic portion of beach that shows changes in terms of profile, morphology and sediment quantity and texture. Changes may be at the scale of daily, diurnal, seasonal or annual. Waves, tides and currents may effect changes.

Aftershock : Earthquakes that follow the largest shock of seismic activity. They are smaller than the "mainshock" in terms of magnitude and intensity. Minor seismic vibrations that precede major shock are termed "foreshock".

Agglomerate : Consolidated pyroclastic deposit composed of volcanic fragments and bombs.

Aggradation : It is the process of accumulation of sediment to cause surface elevation and or morphological change of depositional topography/landscape.

Alluvium : Sediment deposits accumulated during since ice age principally under the influence of rivers along channel course during normal flow and all over flood plain during flooding events. Although sediment size varies as a function of sediment load, carrying capacity and energy conditions, a general decrease is discernible from channel bank to regions far away.

Amplification : Response of earth materials depending on their density to the impending seismic waves. It is dependent on intensity and duration of shaking, topography of the region, geometry and velocity structure of rock types, etc.

Amplitude : The size of the wiggles on an earthquake recording measured in nanometers (10^{-9} m) or micrometers (10^{-6} m).

Angle of repose : The maximum angle at which a pile of unconsolidated material can remain stable under normal gravitational regime. It is a critical measure of soil profile in hill slopes in determining their landslide vulnerability. Various parameters such as vegetation cover, moisture, mineralogy, texture, etc. determine the stability of soil profiles in addition to angle of repose and angle of repose itself.

Anticline : A convex folding in rocks in the shape of inverted "U" in which the strata dip outward on both sides away from the axis. The central part of which contains the oldest section of rock. *Antonym : "Syncline" in which the strata show the shape of "U", dip inward and youngest rocks are found in central part.*

Arc : Commonly refers to the chain of volcanoes (volcanic arc) produced by subduction.

Archipelago : A group of islands located far away from mainland. Genetically the islands may have been formed principally by volcanic activity in combination with other modes of origin.

Aridity index : The ratio of a region's potential annual evaporation, as determined by its receipt of solar radiation, to its average annual precipitation.

Armor : Fixed engineering structures made up of rock or concrete, on or along the shoreline or riverbanks to reduce erosion. Armoring structures include seawalls, revetments, bulkheads, and rip rap (loose boulders).

Array : An ordered arrangement of seismometers or geophones.

Arrival time : The time at which a particular wave phase arrives at a station. The term is used in seismology and oceanography.

Arroyo : A small, deep, usually dry channel eroded by a short-lived or intermittent desert stream.

Artificial headlands : Man- made offshore structures connected to the shoreline to provide coastal protection or to restrict longshore transport.

Ash cloud : A cloud formed from tiny ash particles (less than 2 millimeters in diameter) and gases blasted from a volcano.

Ash flow : A turbulent mixture of hot gas and ash sized rock fragments, ejected from a crater or fissure. The mass of pyroclastics moves rapidly and can affect a large area within short span of time.

Asthenosphere : A layer of soft but solid, mobile rock comprising the lower part of the upper mantle. It is located from about 100 to 350 kilometers beneath the earth's surface. Many of the deep-seated earthquakes and volcanic sources are located from this part of earth's interior.

Attenuation : A decrease in seismic-signal amplitude as waves propagate from the seismic source. Attenuation is caused by geometric spreading of seismic-wave energy and by the absorption and scattering of seismic energy in different earth materials. It is a measure of loss or discipiation of wave energy, resulting in a reduction of wave amplitude.

Aulacogen : A tectonic trough on a craton (shield) bounded by convergent normal faults.

Aureole : A zone surrounding an igneous intrusion in which contact metamorphism of the "country rock" has taken place.

Avalanche : A large mass of material or mixtures of material falling or sliding rapidly under the force of gravity. Avalanches often are classified by their content, such as snow, ice, soil, or rock. A mixture of these materials is a debris avalanche. Velocities may sometimes exceed 500km/hr. They form in regions of active mountain building and steep slopes. Avalanches can be triggered by many factors such as seismicity, volcanism and vibrations created by sking!

Avulsion : 1. Rapid wave erosion in coastal region during a storm. **2.** A sudden cutting-off of land by flood, currents or change in course of river.

Axial plane : A planar (two-dimensional) surface that connects the hinge lines of the strata in a fold.

Azimuth : Horizontal direction to the horizon from a point on the landscape or an observer, measured in 360 degrees of arc clockwise from the north (north=0°, east=90°, south=180°, west=270°). The direction is measured clockwise in degrees against north. It is used to measure the direction from seismic source to a seismic station. It is also utilized in determining paleocurrent directions in older clastic deposits.

Backarc : A linear region located between landwardside of the chain of volcanoes (volcanic arc) and the adjacent continental mainland.

Backshore : The portion of a beach that extends from the hightide line inland to the sea cliff or vegetation line. Swash reaches the backshore only during major storms.

Backswamp : The section of a floodplain located behind natural levees. Backswamps receive sediment influx only during flooding seasons wherein fine detritus get accumulate.

Backwater : The low-lying lands adjacent to a stream that are flooded only during periods of high water. They are located between coastal plain and flood plain. Seasonally they are brought under tidal waters. Mudflats are associated with these areas and together these form fragile ecosystems, acting as nursery for many hundreds of fresh, brackish and saline water fauna and flora, in addition to migratory avian fauna.

Bajada : An alluvial apron. It is a broad sloping area of debris developed by coalescing alluvial fans. These are fertile but environmentally sensitive regions. If not properly stabilized and managed, could become zones of active avalanche/mud/ landslides.

Bank : An elevated embarkment bordering waterbodies.

Bar : Fully or partly submerged mound of sand, gravel, or other unconsolidated material built on the bottom in shallow water by waves and currents. Bars form in offshore regions, in rivers and estuaries. If located parallel to coastline at a distance from it, the bars are called "shoals"; bars at the mouth of river confluence are termed as "distributary mouth bars"; bars in the river are termed as "braided bars". Offshore bars could get welded and form chain of "barrier islands" if they emerge above sealevel. If the raised bars get connected to mainland on one side, the resulting region is termed as "barrier spit". All these geomorphic entities are vulnerable to rising sealevel, storms, cyclones and flood owing to their proximity to water level.

Barchan dune : A crescent-shaped dune that forms around a small patch of vegetation, lies perpendicular to the prevailing wind direction and has a gentle, convex windward slope and a steep, concave leeward slope. Barchan dunes typically form in arid, inland deserts with stable wind direction and relatively little sand.

Basal sliding : The process by which a glacier undergoes thawing at its base, producing a film of water along which the glacier then flows. Basal sliding primarily affects glaciers in warm climates or mid-latitude mountain ranges.

Basalt : A type of volcanic rock rich in iron and magnesium and containing lesser amount of silica than other types of volcanic rocks. It is the most abundant solidified lava on earth's crust. Common rock type ejected from effusive type of volcanoes.

Base flood : Flood that has one percent probability of being equaled or exceeded in any given year (considered in a span of approximately 100 years from the year of computation).

Base Level : The lowest elevation to which a stream is capable of eroding the landscape, usually sea level.

Basement : Igneous and metamorphic rocks that underlie the sedimentary rock sequences of a region and extend downward to the base of the crust.

Basin : An enclosed area or topographic depression that facilitates the accumulation of water and/or sediment.

Bay : A recess or inlet in the shore of a sea or lake between two capes or headlands, not as large as a Gulf but larger than a Cove.

Beach : A deposit of unconsolidated sediment along the edge of a sea that extends landward to the base of the dunes or bluff. It is a dynamic zone that extends landward from the low water line to the place where there is marked change in material or form, or to the line of permanent vegetation (usually the effective limit of storm waves). The seaward limit of a beach is the mean low water line. A beach includes foreshore and backshore.

Beach crest : The point representing the limit of high tide storm wave run-up.

Beach erosion : Removal of beach materials by wave action, tidal and littoral currents or wind.

Beach face : The portion of a foreshore that lies nearest to the sea and regularly receives the swash of breaking waves. The beach face is the steepest part of the foreshore.

Beach nourishment : The process of replenishing a beach, either naturally through accretion, or artificially. Also includes conservation and stabilization measures.

Beach profile : A cross-section taken perpendicular to a given beach contour. An important measure that indicates dynamism of beach process and the effectiveness of beach nourishment efforts.

Beach Ridge : A low extensive ridge of beach material piled up by storm waves landward of the berm. It occurs singly or as a series of more or less parallel ridges. Rows of such beach ridge series are interpreted as paleoshorelines.

Beach scarp : An almost perpendicular slope along the beach foreshore. An erosional feature resulting from wave or longshore current action.

Bedrock : Relatively hard, solid rock that commonly underlies softer rock, sediment, or soil.

Bench : 1. A level or gently sloping erosion plane inclined seaward. **2.** A nearly horizontal area at about the level of maximum high water on the seaside of a dike.

Bench mark : A permanent object in tidal observations, or in a survey, to furnish a datum level.

Benioff zone/ Wadati-Benioff zone : A dipping planar zone experiencing intense and frequent earthquakes and located amidst oceanic and continental crustal plates. Earthquakes originate from this zone as a result of slip faulting, bending and fracturing of plate that goes down due to subduction.

Berm : 1. A nearly horizontal plateau on the beach face or backshore, characterized by sharp break in slope of beach on one side and mainland on other side. It is identified as a linear mound or series of mounds of sediment paralleling the coastline at hightide watermark. **2.** A nearly horizontal area built artificially to support an armour layer. **3.** Earth-filled structures placed on a floodplain to divert flood flows, most commonly into bridge or culvert openings.

Blind fault : A subsurface fault that has no surficial expression. It is "buried" under the uppermost layers of sediments or rocks in the crust.

Blizzard : Winter storms characterized by low temperatures and strong winds that blow large amounts of snow and ice.

Block : A solid rock fragment greater than 64 mm in diameter that was ejected from a volcano during an explosive eruption. Blocks commonly consist of solidified pieces of old lava flows that were part of a volcano's cone.

Block faulting : A type of normal faulting in which the crust is divided into structural blocks of different elevations and orientations; often results in basin-and-ridge topography.

Blowout : Small, often circular or oval depression in sand dunes. The removal of sand from a dune by wind drift after protective dune vegetation has been lost. Unless repaired promptly, the area of blowout will increase in size and could lead to the development of a migrating sand dune and its associated problems.

Bluff : An unvegetated high bank composed largely of unconsolidated deposits with a near-vertical face overlooking a body of water.

Body wave : A seismic wave that propagates through the interior of the earth.

Boil : An upward flow of water in a sandy formation due to an unbalanced hydrostatic pressure resulting from a rise in a nearby stream, or from removing the overburden in making excavations.

Bolide : An asteroid, meteor, comet, or any other solid body moving inside the solar system.

Bomb : Fragment of molten or semi-molten lava, ranging in size upto few feet in dia, blown out during volcanic eruption. Because of their plastic condition, bombs are often modified in shape during their flight or upon impact.

Breaching : Failure of the beachhead or a dike or bank of river or any other waterbody to cause flooding.

Breaker : 1. A wave as it spills, plunges or collapses on a shore. **2.** Natural or man-made structure built for containing outreach of waves.

Breaker zone : The zone within which waves approaching the coastline commence breaking, typically in water depths between 5 m and 10 m.

Breakwater : A wall built seaward of a coast and aligned parallel to it for intercepting incoming waves to protect a harbor or shore.

Brittle-ductile boundary : A depth in the crust across which the thermomechanical properties of the crust change from brittle (above) to ductile (below). Most of the earthquakes in the crust initiate at or above this depth. Below this depth, fault slip may be aseismic.

Buffer Zone : 1. An appropriately managed and unalienated zone of unconsolidated land between beach and development, within which coastline fluctuations and hazards can be accommodated in order to minimise damage to the development. **2.** A zone earmarked for cushioning between protected area and inhabited area.

Bulkhead : 1. A structure separating land and water areas, primarily designed to resist earth pressures. **2.** A structure or partition to retain or prevent sliding of the land. A secondary purpose is to protect the upland against damage from wave action.

Buoy : A float moored to the bottom to mark a channel, anchor, shoal, rock, etc.

Caldera : A vast depression at the top of a volcanic cone, formed when an eruption substantially empties the reservoir of magma beneath the cone's summit. Eventually the summit collapses inward, creating a caldera. Size of caldera may vary from few tens of meters to more than 15 kilometers in diameter and more than 1000 meters deep.

Cape : A relatively extensive land area jutting seaward from a continent or large island which prominently marks a change in, or interrupts notably, the coastal trend.

Catastrophism : The hypothesis that a series of immense, brief, worldwide upheavals changed the Earth's crust greatly and account for the development of mountains, valleys, and other features of the Earth. Also denotes "large scale" destruction caused by geohazards.

Caving : 1. Collapse of a streambank by undercutting due to wearing away of the toe or an erodible soil layer above the toe. **2.** Collapse of vertical or steep cliffs along shorelines due to wave erosion and or undercutting of toe by waves. **3.** Sudden subsidence of land due to collapse of underground karst system.

Check dam : Small sediment storage dams built in the channels of steep gullies to stabilize the channel bed. A common use is to control channelized debris flow frequency and volume. In areas of poor percolation and high surface runoff, check dams are constructed for water storage.

Chronic hazards : A group of hazards that do not stem from one event but arise from continuous conditions (e.g., famine, resource degradation, pollution, and large-scale toxic contamination), which accumulate over time.

Cinder cone : A conical hill formed by the accumulation of cinders and other pyroclastic debris around a volcanic vent.

Circum-Pacific belt : The zone surrounding the Pacific Ocean that is characterized by frequent and strong earthquakes, many volcanoes as well as high tsunami hazard. Also called the "Ring of fire" or "Pacific ring of fire".

Cleavage : Planes of weakness present in specific types of minerals and rocks along which they break easily. It usually refers to flat, reflective surfaces in some minerals, but may also be applied to foliated metamorphic rocks, like slate.

Cliff : A high steep face of rock in hilly regions or bank or coastline at the water's edge. Undercutting of toes due to gravity fall (in hilly regions) or wave erosion (in coastal region) or water flow (along stream) can cause destabilization of adjacent regions.

Coastal defense : General term used to protect coast against erosion and flooding.

Coastal/shoreline management : Policies, strategies, measures and activities undertaken to reclaim, stabilize and nourish coastal region.

Coastal zone : The land-sea-air interface zone around continents and islands extending from the landward edge of a barrier beach or shoreline of coastal bay to the outer extent of the continental shelf.

Coastal erosion : The wearing away of coastal lands by wave attack, tidal or littoral currents or wind or a combination of all these.

Coastline : The line that forms the boundary between land and the sea. Its location is dynamic due to seasonal, tidal & other variations.

Coherent slides : Landslides that consist of a few relatively intact blocks of rock or soil that move together.

Cohesionless : Referring to the condition of sediment grains without bonding between them. Sediment deposits with such condition have no shear strength. As they depend only on friction, they are prone to wind erosion.

Cohesive lahar : A type of volcanic debris flow that contains more than 3-5 percent clay. Such lahars form from volcanic debris avalanches. Cohesive lahars are very mobile and may flow long distances.

Cohesive soil : Microscopic soil particles that have natural resistance to being pulled apart at their point of contact.

Coir log or coir roll : Roll of coconut fiber, used in bioengineering systems to provide erosion control along a streambank and to support the establishment of vegetation.

Colluvium : Loose soil or rock fragments on or at the base of gentle slopes or hillsides. They are generally susceptible to creep or other gravity driven mass wasting processes. Proper stabilization measures have to be undertaken in colluviul deposits lying at or near angle of repose and adjoining settlements, failing which disasters may result.

Conservation : The protection of an area, or particular element within an area, accepting the dynamic nature of the environment.

Continental collision : The convergence of two continental plates, resulting in the formation of mountain ranges. Himalayan mountain chain is the result of collision between Euroasian plate and Indian plate. These regions are prone from seismic, landslide, avalanche and volcanic hazards.

Continental crust : The Earth's crust that includes both the continents and the continental shelf.

Continental drift : The hypothesis, proposed by Alfred Wegener, that today's continents broke off from a single supercontinent and then plowed through the ocean floors into their present positions. This hypothesis emanated first from matching of continental boundaries, similarity of fossil and rock type occurrences in continents located thousands of kilometers away such as India and Africa; Africa and South America, etc.

Contour : A line connecting points of equal value. For example, equal elevation onland (topographic contour) or depth in waterbodies (bathymetric contour).

Contour cropping : Tillage and planting of a crop across, rather than in the direction of, a slope to reduce soil loss.

Contour Interval : The difference in elevation or depth between successive contour lines.

Convergence : The coming together of two lithospheric plates. When one or both the plates are oceanic, subduction takes place. If both the plates are continental, mountains form. In either case, seismicity and volcanism result from these convergence margins.

Core : 1. A cylindrical sample extracted from rock formations below surface. **2.** An inner, often much less permeable portion of a breakwater or barrier beach. **3.** The innermost layer of the Earth. The core is the densest layer of the Earth, and is divided into the outer core, which is believed to be liquid, and the inner core, which is believed to be solid. The outer core extends from about 2900 to about 5120km below the Earth's surface and consists in its main components of a mixture of liquid iron and nickel. The inner core is the central sphere of the Earth with a diameter of 1250km and consists of solid metal.

Country rock : The rock present in any area before some other event affects it, such as faulting or an igneous intrusion.

Cove : A small sheltered recess in a shore or coast, generally inside a larger embayment.

Crater : A small funnel-shaped depression in the summit of a volcano at the top of the conduit or pipe through which the magma reaches the surface.

Craton : A part of the earth's crust that has attained stability and has been little deformed in recent geologic time. It usually refers to the oldest crust of a continent. *Synonym : "shield".*

Creek : A small tidal channel running through a coastal marsh. Flood tide water enters through it, and during ebb tide, it acts as a drain.

Creep : Slow, more or less continuous movement of fault blocks occurring along fault plane. Also applied to slow movement of landslide masses downslope due to gravity.

Critical facilities : Structures whose ongoing performance during an emergency is required or whose failure could threaten many lives. These may include : **1.** Structures such as nuclear power reactors or large dams whose failure might be catastrophic. **2.** Major communication, utility, and transportation systems. **3.** Involuntary or high-occupancy buildings such as schools or prisons; and **4.** Emergency facilities such as hospitals, police and fire stations and disaster-response centers.

Crust : The outermost 5.5 to 65km thick layer of the earth. The uppermost 15-35 km of crust is brittle enough to produce earthquakes. The seismic crust is separated from the lower crust by the brittle-ductile boundary.

Cutbank : The outside bank of a bend, often eroding and across the stream from a point bar.

Cutoff : A channel cut across the neck of a bend.

Cyclone : A low-pressure system. Tropical cyclones originate in the low (tropical) latitudes. During conditions of high surface temperature and moisture over the tropical oceans they evolve into severe tropical storms and hurricanes.

Dam : Structure built in rivers or estuaries, to separate water at both sides and/or to retain water at one side.

Damage potential : The level of damage that could be caused by any hazardous event. It also depends on the degree of susceptibility of a geographic area.

Damping : The reduction in amplitude of a seismic wave or oscillator due to friction and (or) the internal absorption of energy by matter.

Datum : Any position or element in relation to which others are determined, as datum point, datum line and datum plane.

Debris : Solid objects or masses carried by transport agents such as water, wind, ice or gravity.

Debris avalanche : A debris avalanche may be triggered by an eruption or earthquake. Debris avalanches move at velocities ranging from a few tens of meters per second to more than 100 meters per second and behave like complex granular flows or slide flows. Commonly they are voluminous and may run for considerable distances from their source. The resulting debris-avalanche deposit usually exhibits hummocky surface morphology. **1.** Geologic deposit and event related to a landslide, but containing abundant snow and/or ice. As the debris tumbles downslope, the snow and ice break apart giving the slide more mobility and speed. **2.** Rapidly moving, dry flows of disaggregated rock debris, sand, and silt. **3.** A large portion of the volcano may become unstable, break away from the volcanic massif, and become an avalanche. Volcanic debris avalanches commonly form by some type of structural collapse of the volcano, usually the steep front of the cooled lava dome, or other parts of the upper edifice.

Debris basin : A concrete, bowl-shaped pit built on a mountainside to collect and arrest advancing debris flow.

Debris cone : A cone shaped deposit of sediment, usually associated with the accumulation of talus or rock fall or alluvial fans.

Debris flow : Sometimes referred as earth flow or mudflow. **1.** The rapid, downward mass movement of particles coarser than sand, often including boulders, at a rate ranging from 2 to 40 kilometers per hour. Debris flows occur along fairly steep slopes. They are also called lahar. **2.** A type of landslide made up of a mixture of water-saturated rock debris and soil with a consistency similar to wet cement. Debris flows move rapidly downslope under the influence of gravity.

Deflation : The removal of loose material from a beach or other land surface by wind action.

Deflector : Structural barrier (groin, jetty) projecting into a stream or sea to divert flow away from eroding sections of streambank or coast.

Deformation : A change in the original shape of a material or rock formations during tectonic movements and earthquakes.

Degradation : 1. Lowering of the land surface due to erosion, or the process of fluvial erosion and transport of floodplain sediment such that through time the elevation of the floodplain decreases. **2.** Negative change of status of ecology or environment by natural or anthropogenic causes.

Deposition : Accumulation of material by mechanical settling from water or air, glacier, gravity and chemical precipitation from solution by physico-chemical processes and biochemical processes or a combination of all these.

Desert : A region with very low average annual rainfall, sparse vegetation, typically having thin, dry, and crumbly soil. A desert has an aridity index > 4.0.

Desert pavement : A closely packed layer of rock fragments concentrated in a layer along the Earth's surface by the deflation of finer particles.

Desertification : The process through which a desert takes over a formerly non-desert area. When a region begins to undergo desertification, the new conditions typically include a significantly lowered water table, a reduced supply of surface water, increased salinity in natural waters and soils, progressive destruction of native vegetation and an accelerated rate of erosion.

Digital terrain model (DTM) : A Three-dimensional model of digital elevation data for cartographic representation. Terrain models are often displayed as grids, superimposed over topography maps to illustrate peaks and valleys.

Dike, Dyke : 1. A tabular body of igneous rock that cuts across the structure of adjacent rocks or cuts through massive rocks. **2.** A structure designed either to reduce the water velocity as streamflow passes through the dike so that sediment deposition occurs instead of erosion (permeable dike), or to deflect erosive currents away from the streambank (impermeable dike). Such structures are also termed as groin, palisade, spur, jetty or deflector. Dikes along rivers are called levees.

Dip : The amount and direction of tilt of a geologic surface from horizontal, measured in degrees.

Dip Slope : A natural slope of the land surface that relates primarily to the tilt (dip) of the underlying bedrock or tilt of the bedding planes with in a tectonically uplifted block of sedimentary rock. In hilly regions, if the dip direction and topographic slope are on the same direction, then the area is considered to be vulnerable to landslides. Such regions should not be deforested and should not be selected for any developmental activity.

Dip-slip fault : A fault in which two sections of rock have moved apart vertically, parallel to the dip of the fault plane.

Disaster : A singular or interactive hazard event that has a profound impact on local people or places either in terms of loss of life or injuries, property damages, or environmental impact.

Disaster proneness : Measure of the vulnerability of nations or a geographic area to natural disasters. Some of the most disaster-prone countries are those with high hazard frequencies and low national wealth, meaning that they could not respond effectively in the aftermath of a hazard or to mitigate the impact of future hazard events.

Displacement : The difference between the initial position of a reference point and any later position. **1.** In seismology, displacement is the ground motion commonly inferred from a seismogram. For example, it may be calculated by integrating an accelerogram twice with respect to time and is expressed in units of length, such as centimeters. **2.** In geology, displacement is the permanent offset of a geologic or man-made reference point along a fault or a landslide.

Dissection : The erosion of a landscape by various processes that leads to the creation of topographic relief. Often it refers to fluvial erosion creating a series of gullies, valleys, or canyons.

Divergence : The process by which two lithospheric plates separated by rifting and move farther apart, with soft mantle rock rising between them and forming new oceanic lithosphere.

Dormant : An active volcano, which is quiet, not presently erupting, but is expected to erupt in the future.

Downthrown : Side/part of a fault that appears to have moved downward compared with the other side. *Antonym* : "upthrown".

Drainage basin : Total area drained by a stream and its tributaries.

Drawdown : Lowering of water levels in rivers, lakes, or underground aquifers due to pumping or artesian flow. Drawdown may destabilize unsupported banks or poorly packed earth to sink (subsidence) and also can cause landslides along hillslopes.

Drought : Extended period of unusually dry weather that results in a shortage of water and consequently often a decrease in or failure of food crops.

Dune : Mound or ridge of sand deposited by the wind, capable of movement when unvegatated. Sand fencing or plantation can enhance dune building.

Dune management : The general term describing all activities associated with the restoration and/or maintenance of the role and values of beach dune systems.

Earth fissure : A tension fracture in the Earth's surface that results from substantial overdrafting (pumping) of ground water and subsequent subsidence of the ground surface. Subsidence and resultant earth fissures may occur as a result of dissolution of soluble rock formations present underground.

Earth systems : The Earth is made up of four basic systems : the lithosphere (the rocks of the Earth); the hydrosphere (the waters of the Earth); the atmosphere (the gases that surround the Earth); and the biosphere (the life on Earth). These systems interact to produce most of the geological processes that occur on Earth. An event involving one of these systems may affect some or all of the others.

Earthquake : Ground shaking and radiated seismic energy caused by sudden slip on a fault, volcanic or magmatic activity, or other sudden stress changes in the earth. An earthquake is measured commonly on Richster scale of 1-10 in which successive increment is on logrithemic scale; i.e., a progression from 1 to 2 is measured such that the scale 2 is 10 times intense than magnitude of 1 and so on.

Earthquake hazard : Any phenomenon associated with an earthquake that may produce adverse effects on environment, lives and property. It is defined as the probability of occurrence of a specified level of ground shaking in a specified period of time.

Effusive eruption : An eruption dominated by the outpouring of lava onto the ground is termed effusive eruption (as opposed to the violent fragmentation of magma by explosive eruptions). Lava flows generated by effusive eruptions vary in shape, thickness, length, and width depending on the type of lava erupted, discharge, slope of the ground over which the lava travels, and duration of eruption. Deccan pleatue of India is made up of solidified lava flow from effusive eruption that lasted for over 3.5 million years. It is considered to be one of the greatest volcanic events of Earth's history.

Ejecta : Material that is thrown out by a volcano.

Emergency spillway : A channel other structure constructed at a dam to pass flood flows that exceed the design capacity of the flow through the said structure.

Emergent coast : A coast in which land formerly under water has recently been placed above sea level, either by uplift of the land or by a drop in sea level or by aggradation or by a combination of few of these or all of these processes.

Energy dissipater : A rock or concrete structure designed to reduce the velocity of the flow exiting a culvert to prevent erosion of the streambed and banks.

Epicenter : The point on the earth's surface vertically above the point (focus or hypocenter) in the crust where a seismic rupture nucleates. This point is expressed by its geographical latitude and longitude.

Epidemic : Rapid outbreak of a disease, affecting hundreds or thousands of people.

Erosion : The process of wearing away of landmasses, generally involving the detachment and movement of sediment and rock fragments through the action of wind or water or any other agent. The rate of erosion is dependent on the resistance of rocks, intensity and period of erosional agents.

Eruption : The expelling of material including gases, ash, volcanic fragments and lava on Earth's surface due to volcanic activity. Eruptions may be explosive or effusive.

Escarpment : A cliff, bluff, or steep slope. Substantial topographic relief over a short distance across the land surface often associated with tectonic uplift along a fault or intensive fluvial erosion/ downcutting.

Estuary : Estuary is an important ecosystem, which supports a variety of physico-chemical interaction between fresh water and saline water and promotes sedimentation. Unique fauna and flora flourish in these regions that are tolerant to salinity and turbidity. Any ecological disturbance in this region would devastate upstream environment as well as deep marine conditions. **1.** A semi-enclosed coastal body of water, which has a free connection with the open sea. The seawater is usually measurably diluted with freshwater. **2.** The landward extension of seawater through a river channel. The maximum distance from coastline upto which the tidal fluctuations are observed, defines upper limit and type of estuary. Considering constant sealevel, downward movement of this upperlimit is considered to be progradational nature of the delta while upward movement is indicative of destructive phase of delta.

Extrusive rock : An igneous rock formed from lava that has flowed out onto the Earth's surface, characterized by rapid solidification.

Fallout : A general term for debris that falls on the earth from an eruption cloud.

Famine : State of massive food deprivation leading to malnutrition and death of large number of people. Famine may result from drought as well as lack of critical facilities for relief and rescue operations conducted during the events of geohazards.

Fault : A fracture along which there has been significant physical/ geographic/altitudinal displacement of the two sides relative to each other parallel to the fracture. Strike-slip faults are vertical or near vertical fractures along which rock masses have mostly shifted horizontally. Dip-slip faults are inclined fractures along which rock masses have shifted vertically. If the rock mass above an inclined fault is depressed by slip, the fault is termed normal, whereas if the rock above the fault is elevated by slip, the fault is termed reverse (or thrust). Oblique-slip faults have significant components of both slip styles.

Flash flood : A sudden-onset of severe, but often locally restricted flood event occurring mainly in arid or semi-arid areas with steep topography following intense, short-lived rainstorms. Also occur in narrow valleys or developed urban areas where impervious surfaces increase rapid runoff. Lag time between onset of syndrome and actual impact is very limited and thus this hazard is highly devastating, getting the people unwares, often resulting in huge number of fatalities.

Flood : Either a general and temporary condition or partial or complete inundation of normally dry land areas from the overflow of inland or tidal waters caused by the unusual and rapid accumulation or runoff of surface waters from any source. Coastal flooding results when water surpasses its tide line. Flood problems can be exacerbated by human activities such as land drainage, river channelization, unplanned construction and deforestation.

Foreshore : The position of beach that extends from low tide line up to back shose.

Fumarole : An opening located in geothermal regions through which superheated gas and steam are discharged.

Gabion basket : A type of material made up of metal wire box containing ripraps of rock fragments. The boxes are made in the shape of bricks and are stacked in regions where defense mechanisms are required. These baskets allow percolation of water and deposition of fine detritus to fill-in the gaps present between ripraps within the basket. In due course of time, the metal wire rusts away, forming a naturally cemented erosion/weathering resistant defense.

Geographic information system (GIS) : A type of MIS (Management Information System) with the capabilities of storing large spatial and alphanumeric data and inbuilt query options. It differs from conventional MIS in such a way that the results are displayed in spatially referenced geographic details. It has ability to analyze, manipulate and display geographic data. Decision support systems, dynamic modeling of geological processes and estimation of extent of damage are all made with ease through GIS.

Geohazard : A geological state of natural process that endangers lives and property.

Geomembrane : An impervious thin sheet of rubber or synthetic material that functions as a barrier for liquid or vapor.

Geotechnics : The use of engineering principles to understand how earth materials, such as soils and rocks, behave. This knowledge is applied to design engineering structures in order to curtail damage causing potential of natural processes.

Geotextile : Textile made from synthetic fibers, usually nonbiodegradable. Geotextiles can be woven or nonwoven and varying degrees of porosity. They are used as moisture barriers, for separation or reinforcement of soils, for filtration and for drainage.

Groin : Artificial construction structure located parallel to the shoreline for reducing littrolal transport and coastal erosion. They are designed to reduce wave and current speed, promoting decelaratin and deposition of sediments carried by them.

Gully : Deep channel running for short distances. It is created by excessive surface run-off and resultant erosion of ground surface. It causes headland erosion and clogging of downstream regions through excessive sediment supply.

Guyot : A flat seamount, denuded by the action of waves and streams.

Hazard : Any event or a process that is detrimental to the safety of people and their properties.

Hinterland : The land region located beyond highest hightide line along the coast.

Hydrothermal vent : A submarine opening that discharges warm to superhot mineral rich water. It is usually located on the seafloor associated with spreading centers and or subsurface magma chambers.

Hypocenter : The sub-surface point from where earthquakes are probagated. It is also termed as "focus".

IMD : India Meteorological Organization. It is responsible for monitoring weather conditions and forecast monsoons, cyclones and other atmosphere-born natural hazards.

Intrusion : Penetration of magma into preexisting rock. If it reaches the surface, then the process is called "extrusion".

Island arc : A curved chain of volcanic islands rising from the deep-sea floor and near to a continent caused by subduction processes and occurring on the continent side of the subduction zone. Its curve generally is convex toward the open ocean.

Lahar : Debris flow containing angular clasts of volcanic material. The term had originated from Indonesia.

Landform : External morphology of land surface. Also called as geomorphic feature. Earth's processes shape land surface into different shapes and thus study of landforms (geomorphology) documents past events that operated in an area, helping to deduce stability/suitability/vulnerability of the region.

Landslide : Downward movement of land or soil or rock under the influence of gravity, but aided by many other factors such as water, hydrostatic pressure, slope, etc.

Lapilli : Solidified lava fragments ejected from volcana by explosion and ranging in size between 2 and 64mm.

Lateral Blast : A relatively rare explosion of hot, low-density mixture of rock debris, ash, and gases that move at high speed from outside of the vent (laterally) rather than up from the vent (vertically).

Lava : When magma (molten rock) erupts onto the earth's surface, it is termed as lava.

Leeward : A term used in describing dunes. It is the direction of blowing of wind. Opposite direction of windward side of a dune. It is also used to indicate the direction of wave travel.

Liquefaction : It is a process by which water-saturated sediment temporarily loses strength and acts as a fluid. Vibrations from earthquake can cause this effect. Liquefaction in porus rocks and soils that contain water or under pore pressure on a slope can trigger landslide.

Lithosphere : The outer solid part of the earth that includes the crust and uppermost mantle. The lithosphere is about 100 km thick, although its thickness is age dependent.

Littoral currents : A current running parallel to the beach and generally caused by waves striking the shore at an angle.

Littoral drift : Movement of sand by littoral currents in a direction parallel to the beach along the shore.

Littoral Zone : The littoral zone extends from the onshore dune system to the seaward limit of the offshore zone and possibly beyond. It is the area of the coastline in which sediment movement by wave, current and wind action is prevalent.

Love wave : A major type of seismic surface waves having a horizontal motion that is transverse (or perpendicular) to the direction of propagation. It is named after A. E. H. Love, the mathematician who discovered it.

Low tide : Minimum elevation reached by each falling tide. Also referred as ebb tide.

Low-velocity zone : An area within the Earth's upper mantle in which both P waves and S waves travel at markedly slower velocities than in the outermost part of the upper mantle. The low-velocity zone occurs in the range between 100 and 350 kilometers of depth.

Lunar tide : The portion of the tide that can be attributed directly to attraction to the Moon.

Maar Volcano : A low-lying, broad volcanic crater formed by multiple shallow explosive eruptions. It is more or less circular and often contains a lake, pond or marsh.

Magma chamber : The subterranean cavity containing the gas-rich liquid magma which feeds volcano(es).

Magnitude - earthquake : A number that characterizes the relative size of an earthquake. The Richter scale measures the magnitude of earthquakes through the amount of energy released. The Mercalli's scale measures the intensity in terms of damage caused by an earthquake.

Magnitude - geohazards : The quantum of damage caused by any geohazard is termed as magnitude.

Mainshock : The largest earthquake in a sequence, sometimes preceeded by one or more foreshocks and almost always followed by many aftershocks.

Mantle : The middle layer of the Earth, lying just below the crust and consisting of relatively dense rocks. The mantle is divided into two sections, the upper mantle (SIAL) containing predominantly silica and alumina rich rock material and the lower mantle (SIMA) containing predominantly silica and magnesium rich rock material.

Mass wasting : A general term for the downslope movement of soil and rock material under the influence of gravity. Rates (velocities) of mass wasting phenomenon range dramatically from centimeters per year to 160km/hour.

Mean sea level (MSL) : Average height of the sea surface over a 19-year period, usually determined from hourly height readings.

Meander : A geomorphic feature of a stream at its old stage shown in the form of a curved loope and formed in flood plains only. When the flow conditions in the stream change cross cutting of meander and straightening of the channel course occurs. Isolated such loops form oxbow lakes. While the outer curve of the meander constantly shifts further up through erosion, the inner curve always deposits sediments to form point bars and shifts towards the outer curve. While developmental activities located along outer curve are prone to flooding and erosion, the inner curves are prone to flooding if not levees are formed either naturally or through artificial means.

Melange : A body of rocks consisting of large blocks (mappable size) of different rocks jumbled together with little continuity of contacts.

Mercalli intensity scale : A scale designed to measure the degree of intensity of earthquakes, ranging from I for the lowest intensity to XII for the highest. The classifications are based on human perceptions on the damage caused by an earthquake event.

Microearthquake : Earthquake recognizable only by seismograph and not perceptible by humans. Generally, earthquakes of magnitude less than 2 in Richter's scale are termed as microearthquakes.

Microseism : A more or less continuous motion in the Earth in a wide frequency range that is unrelated to any earthquake and caused by a variety of usually uncorrelated (incoherent) natural and artificial (man-made) sources. Seismic noise or microseism are generated by wave motions on lakes and oceans and their action on shores, typically with periods between about 2 to 9 (the stronger secondary microseisms), and 11 to 18 seconds (the weaker primary microseisms).

Microzonation : Designation of least measurable areas vulnerable to particular geohazard based on local geological, geographic, demographic and climatological conditions.

Mid-oceanic ridges : Elongated rises on the ocean floor where basalt periodically erupts, forming new oceanic crust.

Mitigation : Measures or actions that lessen the harmful effects of disasters and hazards. It includes all those activities directed towards eliminating or reducing the probability of occurrence of a disaster-producing event, or reducing the effects of those events that are unavoidable.

Moho : A discontinuity in seismic velocity that defines the boundary between crust and mantle of the Earth. Also termed the mohorovicic' discontinuity after the Croatian seismologist Andrija mohorovicic' (1857-1936) who had discovered it. It lies between 25 and 60 km deep beneath the continents and between 5 and 8 km deep beneath the ocean floor.

Moment-magnitude scale : A recently developed alternative to the Richter scale used to measure more accurately the amount of energy released by large earthquakes.

Monitoring system : A system for monitoring geohazards such as earthquakes, volcanic eruptions, tsunami, and/or other phenomena, consisting of a network of sensors of various types depending on the type of hazard. In recent times, the system is inclusive of those entire precautionary infrastructure, command and control structures and the protocols of relief and rescue operations.

Moraine : A mound or ridge of unstratified glacial debris deposited by the direct action of glacial ice. Depending on their location with reference to the glacier, they are termed as lateral moraines, terminal moraines, etc.

Morphodynamics : Change of external morphology of land surface by earth's processes such as stream current, waves, tectonics, weathering, etc.

Mud volcano : A mud volcano is a cone of mud and clay, usually less than 1-2m tall. It is built by a mixture of hot water and fine sediment (mud and clay) that either pour gently from a vent in the ground like a fluid lava flow or eject into the air like a lava fountain by escaping volcanic gas and boiling water.

Mudflow : A general term for a mass-movement landform and process characterized by a flowing mass of fine-grained earth material with a high degree of fluidity.

Mulch : Any material used to cover a soil surface to reduce evaporation or prevent erosion.

Normal fault : A fault in which vertical motion has occurred where the hanging wall moved down relative to footwall. It is also termed as gravity fault.

Nourishment - beach : The process of replenishing a beach, either naturally through accretion due to the longshore transport, or artificially through addition of sandy materials.

Nourishment - general : Preservation and mitigation measures that attempt to safeguard ecological or environmental status of an area.

Oblique slip : Fault motion that involves both dip-slip and strike-slip movements of fault blocks.

Obsidian : A black or dark-colored volcanic glass, formed by rapid cooling and thus lacking crystallographic textures (amorphous), usually of rhyolite composition, characterized by conchoidal (curved) fractures.

Oceanic spreading ridge : A fracture zone along the ocean bottom that accommodates upwelling of mantle material to the surface, thus creating new crust. This fracture is topographically marked by a line of ridges formed by solidification of extruded molten magma at the ocean bottom.

Oceanic trench : Deep steep-sided depression in the ocean floor caused by the subduction of oceanic crust beneath either other oceanic crust or continental crust. Mariana trench located in Pacific Ocean is considered to be the deepest part of Ocean all over the World.

Orogeny : The process of formation of mountains, including thrusting, folding and faulting in the outer and higher layers and plastic deformation, metamorphism and plutonism in the inner and deeper layers.

Oxbow lake : A crescent-shaped body of standing water formed from a single loop that was cut off from a meandering stream. *Also see* "Meander".

P wave : Short form for Primary Wave. P stands for "unda prima" = primary wave. A seismic body wave that involves particle motion (alternating compression and extension) in the direction of propagation. P waves travel faster than S waves and, therefore, arrive earlier in the record of a seismic event.

Pahoehoe : A Hawaiian term for basaltic lava flows typified by a smooth, billowy or ropy surface.

Passive continental margin : A border that lies between continental and oceanic lithosphere, but is not a plate margin. It is marked by lack of seismic and volcanic activity.

Peak ground acceleration (PGA) : Maximum acceleration amplitude measured or expected in a strong-motion accelerometer record of an earthquake.

Peak period : The wave period determined by the inverse of the frequency at which the wave energy spectrum reaches it's maximum.

Pele hair : A natural spun glass formed by blowing-out during quiet fountaining of fluid lava, cascading lava falls, or turbulent flows, sometimes in association with Pele tears. A single strand, with a diameter of less than half a millimeter, may be as long as two meters.

Pele tears : Small, solidified drops of volcanic glass behind which trail pendants of Pele hair. They may be tear-shaped, spherical or nearly cylindrical.

Peneplain : A nearly flat erosional surface, usually near sea level, produced by the long-continued processes of mass wasting, sheetwash, and stream erosion.

Peninsula : An elongated portion of land surrounded by water and connected to a larger body of land. Indian sub-continent is considered to be a peninsula.

Permafrost : Any soil, subsoil, or other surficial deposit or even bedrock, occuring in the arctic, subarctic, and alpine regions at variable depth beneath the Earth's surface in which a temperature below freezing has existed continuously for a long time is termed permafrost. This definition is based exclusively on temperature regardless of texture, degree of compaction, water content, and lithologic character of the material.

Permeability : The capability of a given substance to allow the passage of a fluid. Permeability depends on the size and the degree of connection among a substance's pores.

Phreatic eruption : An explosive volcanic eruption caused when water and heated volcanic rocks interact to produce a violent expulsion of steam and pulverized rocks.

Pier : Open construction structure, extending out into the water from the shore, to serve as a landing place, recreational facility, etc.

Pillow Basalt : A particular type of the volcanic rock basalt that forms pillow-shaped masses, indicative of their underwater formation.

Plate tectonics : It is the scientific theory that postulated that the earth's outer shell is composed of several large, thin, relatively strong "plates" that move relative to one another. Movements on the faults that define plate boundaries produce most earthquakes.

Plinian Eruption : Plinian eruptions are one of the most explosive types of eruptions, forming enormous dark columns of tephra and gas high into the stratosphere.

Plug : Solidified lava that fills the conduit of a volcano. It is usually more resistant to erosion than the material making up the surrounding cone and may remain standing as a solitary pinnacle when the rest of the original structure has eroded away.

Plug dome : When viscous lava wells up into a crater and is too stiff to flow away, it piles up as a dome-shaped mass, often completely filling the vent from which it emerged. The resultant steep-sided, rounded mound is called plug dome.

Plunging Waves : Waves whose crests break suddenly with tremendous force by curling over near vertical faces.

Pluton : Any body of igneous rock that solidified below the earth's surface.

Pluvial lake : A lake that formed from rainwater falling into a landlocked basin during a glacial period marked by greater precipitation than is found in the region in prior or subsequent periods.

Pocket Beaches : Small beach systems bounded by rocky headlands.

Point bar : A stream bar deposited on the inside of a curve in the stream, where the water velocity is low. *See also* "Meander".

Polarity : In seismology, the direction of first motion on a seismogram, either up (compression) or down (dilatation or relaxation).

Polarity reversal : The occurrence of waveforms that are mirror images of their related initial phases, e.g., the waveforms of depth phases with respect to their primary P- or S-type phases.

Porosity : Percentage of pores in soil, rock or sediment. When connected, they define permeability. Permeable pores when filled with water and present along slopes create pore pressure and make the slopes vulnerable to downsliding.

Preparedness : The degree of alertness and readiness of an individual or a community immediately before the onset of a hazard event. Preparedness activities are designed to help save lives and minimize damage by preparing people to respond appropriately.

Preservation : Static protection of an area or element, attempting to perpetuate the existence of a given 'state'.

Prevention : Measures aimed at averting hazards before they occur.

Pyroclastic : Rocks formed by volcanic explosion are termed as pyroclastics. Pyroclastic as a verb defines volcanic eruption.

Pyroclastic cone : Steep, conical volcano composed almost entirely of an accumulation of loose pyroclastic material.

Pyroclastic eruption : A volcanic eruption of viscous, gas-rich magma. Pyroclastic eruptions tend to produce solid volcanic fragments rather than fluid lava.

Pyroclastic flow : A dense, hot, chaotic avalanche of rock fragments, gas, and ash that travels rapidly away from an explosive eruption column, down the flanks of the volcano (synonymous with 'ash flow'). Pyroclastic flows move at speeds ranging from 10 to several hundred meters per second and are typically at temperatures between 300° and 800°c. Pyroclastic flows form either by collapse of the eruption column, or by failure of the front of a cooling lava dome. Once these flows are initiated, they may travel several kilometers and easily override topographic obstacles in the flow path.

Pyroclastic surge : A low-density, turbulent flow of fine-grained volcanic rock debris and hot gas. Pyroclastic surges differ from pyroclastic flows in that they are less dense and tend to travel as a low, ground-hugging, but are highly mobile clouds. Surges often affect areas beyond the limits of pyroclastic flows.

Quick clay : Sediment that sets off a sudden mudflow by changing rapidly from solid to liquid form, after an earthquake, explosion or thunder.

Quicksand : A mass of loose, sediment that becomes fluidized when suddenly vibrated.

Quiescent : A volcano, which is not active, but is still registering seismic activity. When there is no more seismicity, the volcano is dormant, but still capable of erupting.

Raised beach : A coastal landform located above the present sealevel, usually expressed as a wave-cut platform of beach or cliff. It is formed due to a combination of processes such as sealevel lowering, coastline retreat towards offshore, tectonic uplift, etc.

Relief : Difference of elevation of a geographic location from a datum plane, usually from MSL.

Reclamation : Retrival of an area from environmental impact/pollution or damage caused by geohazard.

Recovery : The process of returning to "normality" after an emergency occurred.

Recurrence interval : The average time span between events such as flooding or cyclones or earthquakes at a particular site. Also termed as "return period".

Resilience : The ability to wither away the impacts of a hazard event.

Richter scale : A logarithmic scale that measures the amount of energy released during an earthquake on the basis of the amplitude of the highest peak recorded on a seismogram.

Rift : A long, narrow crack in the entire thickness of the Earth's crust, which is bounded by normal faults on either side. Rifting occurs when the brocken pieces of crust are pulled apart.

Rill : A tiny drainage channel cut in a slope by the flow of water. With continued deepening and lengthening, transforms itself into gully.

Ring of fire : The regions of mountain-building earthquakes and volcanoes, which surround the Pacific Ocean.

Rip rap : Irregular shaped broken stone pieces of varying sizes and placed on sites of active beach erosion or along erosional banks of river channel, primarily aimed to deccipate energy and thereby to reduce erosion.

Risk analysis : Assessment of the total risk due to a particular geohazard at a place. Iterations of various magnitudes of different geohazards according to local conditions are performed, based on which hazard preparedness measures are taken for the most likely possibility.

Rock fall : The relatively free falling or precipitous movement of a newly detached segment of bedrock of any size from a cliff or steep slope.

Rubble - natural : Loose, angular, waterworn stones along a beach.

Rubble - general : Rough, irregular pieces of large fragments of rock or concrete.

Runoff : Overland flow of water in the form of sheetwash. It occurs during rainfall exceeding rate of percolation and becomes the cause of erosion.

Run-up height : The maximum height of a tsunami wave when running into shallow waters near the coast.

Rupture zone : The area of the earth through which fault movement occurred during an earthquake. Ruptures may or may not extend to the ground surface.

S wave : A seismic body wave that involves a shearing motion in a direction perpendicular to the direction of propagation.

Saffir-Simpson Hurricane Scale : Measure of hurricane intensity and magnitude by scaling the storm based on central pressure, wind speed, storm surge, and potential damage. Ranges from 1 (minimal) to 5 (catastrophic).

Sand boil : Sand and water ejected to the ground surface as a result of liquefaction at shallow depth. The conical sediment deposit formed due to this process is evidence to it.

Sand Dunes : Mounds or hills of loose sandy sediments formed in arid regions and coastal tracts above beach berm.

Scarp : A low concave cliff or series of cliffs that marks the detachment zone of a landslide or slope failure.

Scoria : A pyroclastic rock composed of vesicular cinders of basaltic composition. Its cellular nature is due to the escape of volcanic gases before solidification.

Sea-floor spreading : The process of adding to the Earth's crust at mid-ocean ridges as magma wells up and forces previously formed crust apart.

Seamount : A conical underwater volcanic mountain rising 1000m or more from the seafloor.

Seawall : A wall of stone, concrete or other sturdy material, built along the shoreline to prevent erosion.

Secondary hazards : Hazards resulting from other hazard events. For example, earthquakes destabilize mountain slopes to cause landslides. Similarly, flooding in cities causes drainage systems to clog, spreading health hazards.

Seiche : A free or standing-wave oscillation of the surface of water in an enclosed or semi-enclosed basin (as a lake, bay, or harbor) that is initiated by local changes in atmospheric pressure, aided by winds, tidal currents and small earthquakes; and that continues in a pendulum fashion, for a time after cessation of the originating force.

Seismic hazard : The potential for damaging effects caused by earthquakes. The level of hazard depends on the magnitude of likely quakes, the distance from the focus and the type of ground materials at a site.

Seismic moment : A measure of the size of an earthquake based on the area of fault rupture, the average amount of slip and the force that was required to overcome the friction of blocks of rocks that were offset by faulting. Seismic moment can also be calculated from the amplitude spectra of seismic waves.

Seismic risk : The chance of injury, damage or loss resulting from seismic hazards. There is no risk, even in a region of high seismic hazard, if there are no people or property that could be injured or damaged by a quake. On the contrary, even a minor earthquake could cause havoc if a region is thickly populated and contains poorly constructed structures.

Seismic swarm : A series of earthquakes, occurring in a limited area over a relatively short period of time.

Seismic wave : An elastic wave generated by an impulse such as an earthquake or an explosion. Seismic waves travel either along or near the Earth's surface (Rayleigh and Love waves) or through the Earth's interior (P and S body waves).

Seismic zonation : Geographic delineation of areas having different potentials for hazardous effects from future earthquakes. Seismic zonation can be done at any scale - national, regional, local or site, the latter two are referred as microzonation.

Seismicity : The geographic and historical distribution of earthquakes.

Seismogenic : Capable of generating earthquakes. Usually refers to earth's processes such as volcanic explosion or landslide that cause the earth to vibrate.

Seismogram : A record written by a seismograph in response to ground motions produced by an earthquake, explosion or other sources that generate oscillating ground motions.

Seismograph : An instrument used to measure the shaking caused by an earthquake.

Shatter cone : A funnel-shaped fracture in a rock (in three dimensions) caused by a sudden release of energy nearby. The tip of the funnel shape points in the direction of the location of energy release.

Shear : Force parallel to a surface as opposed to directly on the surface.

Shear Strength : The capacity for a soil/rock to keep its structural integrity intact under the influence of applied loading and confining stresses. In fine-grained soils, the applied stresses are transmitted to the shear bonds between the soil particles. The strength of these bonds is known as the shear strength of the soil. If the applied load is too great the bonds will break and the soil will fail causing landslides.

Shear stress : The stress component parallel to a given surface, such as a fault plane, that results from forces applied parallel to the surface and transmitted through the surrounding rock.

Shear Zone : A tabular zone of rock that has been crushed by many parallel fractures, typical of a fault zone, due to shear strain.

Sheet erosion : The removal of a fairly uniform layer of soil from the land surface by runoff or wind.

Sheet flow : Movement of sediment grains under high shear stress as a layer that extends from the bed surface to some distance below (on the order of a few cm). Grains are transported in the direction of fluid flow.

Shield volcano : A low, broad, gently sloping, dome-shaped structure that forms over time as repeated eruptions eject basaltic lava through one or more vents and the lava solidifies in approximately the same volume all around.

Shingle : Unconsolidated and unsorted coarse beach material containing a mixture of well rounded gravel, pebble and larger material.

Shore : Narrow strip of land in immediate contact with the sea, including the zone between high and low water lines. A shore of unconsolidated material is usually called a beach.

Shoreline : Intersection of a specified plane of water with the shore or beach.

Sill - sedimentary : Low offshore barrier structure whose crest is usually submerged, designed to retain sand on its landward side.

Sill – igneous : A tabular igneous intrusion parallel to the layering in the country rock, which it intrudes.

Sinkhole : A natural depression in the surface of the land caused by the collapse of the roof of a cavern or subterranean passage, generally occurring in limestone regions.

Sinter : A mineral crust or deposit formed from the minerals (mainly silica) in geothermal water, especially from geysers.

Slickensides : Polished and or striated rock surfaces caused by frictional movement of fault blocks against each other. The intense pressure and temperature generated by the fault movement partially melts and or deforms the surface of fault blocks along the sliding plane, causing striations and polishing, which together are called slickensides. Presence of slickensides is interpreted as evidences of fault movements.

Slide : The mass movement of a single, intact mass of rock, soil or unconsolidated material along a weak plane, such as a fault, fracture or bedding plane. A slide may involve as little as a minor displacement of soil or as much as the displacement of an entire mountainside.

Slip : The relative displacement of formerly adjacent points on opposite sides of a fault, measured along the fault plane.

Slope or Gradient : Degree of inclination to the horizontal. It is expressed as a ratio, such as 1 :25 or 1 on 25, indicating 1-unit vertical rise in 25 units of horizontal distance; or in degrees from horizontal.

Slope erosion : The wearing away of rock or soil from a slope crest, face, or toe due to water, wind, ice or gravity.

Slope instability : It is a state of lack of firmness of earth's surface, primarily found on hilly regions but could be observed along cliff coasts and at the offshore end of continental shelf. Slope instability is caused by cohisionless nature of soil or sediment or a mixure of sediment and rock fragments. Slope instability on hills can cause landslides and avalanches. In submarine conditions, it causes mass wasting and deposition of turbidites.

Solifluction : A form of creep in which soil flows downslope at a rate of 0.5 to 15cm/year. Solifluction occurs in relatively cold regions when the brief warmth of summer thaws only the upper of regolith, which becomes waterlogged because the underlying ground remains frozen and therefore the water cannot drain down into it.

Spatial dispersion : Refers to the pattern of distribution of a hazard or pollutant over a geographic area.

Spatter cone : A low, steep-sided cone of spatter built up on a fissure or vent. It is usually of basaltic material.

Steam (Phreatic) eruption : An explosive volcanic eruption caused when water and heated volcanic rocks interact to produce a violent expulsion of steam and pulverized rocks.

Storm surge : Rise of waterlevel above normal levels along the open coastal regions due to various reasons including cyclones. The reduction of atmospheric pressure in regions above coastal zone can also cause sealevel to rise and inundate large swathes of coastal plains.

Stratovolcano : A large, steep-sided, symmetrical cone built of alternating layers of lava, ash, cinders, blocks, and bombs. Also called composite volcano.

Strike : The direction or trend of a bedding plane or fault, as it intersects the horizontal.

Strike-slip fault : A vertical fault along which the two sides move horizontally past each other. Strike-slip faults are caused by shearing stress.

Strombolian eruption : Intermittent and short duration eruptions ejecting blocks, bombs and lava flows. The name has been drived from the volcano Stromboli.

Subduction zone : The zone of convergence of two tectonic plates, one of which usually overrides the other.

Subsidence : The sudden or slow sinking of the Earth's surface with little or no horizontal motion. The movement is not restricted in rate, magnitude or area involved. Subsidence may be caused by natural geologic processes, such as solution, thawing, compaction, slow crustal warping, or withdrawal of fluid lava from beneath a solid crust; or by man's activity, such as subsurface mining or the pumping of oil or ground water or sudden increase of overburden such as dumping mine tailings or establishment of large constructional structures over clayey soils.

Summit : The top of a mountain or volcano.

Sustainable development : It refers to measures taken to protect natural ecology, economy and social needs. Maintanance of a balance among these three components is considered to be robust developmental activity.

Suture : A complex tectonic structure formed in areas where plate-tectonic convergence of two or more continents take/took place. The rocks and sediments caught up in the suturing event are highly folded, thrust-faulted. In the core of the resulting orogen, the rocks are metamorphosed and selectively melted.

Suture zone : The area where two or more continental plates have joined together through continental collision. Suture zones are marked by extremely high mountain ranges, such as the Himalayas and the Alps and are the regions of seismicity and landslides.

S-wave shadow zone : The region within an arc of 154° directly opposite an earthquake's epicenter that is marked by the absence of S waves. The S-wave shadow zone is due to the fact that S waves cannot penetrate the liquid outer core.

Swell : Wind-generated waves travelling out of their point of origin. Swell characteristically exhibits a more regular and longer period and has flatter crests than waves within their fetch.

Swelling soils : Soils or soft bedrock, which increase in volume as they get wet and shrink as they dry out. The soils rich in clay minerals such as vermiculite, bentonite and montmorillinite show such swelling property.

Syncline : A fold of rock layers that is convex downwards. *Antonym* : "anticline".

Tectonic : Refers to rock-deforming processes and resulting structures that occur in the lithosphere.

Tectonic forces : Forces generated from within the earth that result in uplift, movement or deformation of part of the earth's crust.

Tectonic plates : A rigid section of Earth's crust that moves relative to other such sections on the Earth's surface.

Teleseism : A seismic source that is thousands of kilometers distant from the recording station.

Temporal spacing : Refers to the sequencing and seasonality of events. Some events are quite random (volcanoes) while others have seasons (hurricanes, storm surges and floods).

Tephra : Rock fragment forcibly ejected from the volcano during an eruption.

Terrace : Flat, elevated surface features within or adjacent to the active floodplain that consists of remnants of former floodplains or planar surfaces of fluvial erosion. The stream leaves these features stranded above the floodplain during periods of active incision or downcutting. They are most prominently preserved when the rate of downcutting exceeds the rate of lateral migration and erosion of the stream channel across its floodplain.

Terrain : A tract or region of the earth's surface considered as a physical feature.

Terrane : A general term used to refer to a piece of the crust that is usually smaller than a continent but larger than an island.

Thalweg : A term used to designate the longitudinal profile of a river from source to mouth following the line of the lowest points of a valley.

Thermokarst : Karst-like topographic features produced in a permafrost region by the local melting of ground ice and the subsequent settling of the ground.

Thrust Fault : Earthquake fault formed by the movement of one rock mass over another rock mass at a low angle (45° or less) for most of its extent due to compressional forces.

Thunderstorm : Sudden electrical discharges manifested by a flash of light (lightning) and a sharp rumbling sound. Thunderstorms are associated with convective clouds and are accompanied by heavy rainfall. They are usually shortlived and hit only a small area.

Tidal wave : Marine waves produced by tides and tidal currents. Storm surges and tsunamis are often wrongly termed as tidal waves. Their genetic cause is different and should not be termed so.

Tide : Periodic rising and falling of sealevel resulting from gravitational attraction of the moon, sun and other astronomical bodies acting upon the rotating earth.

Tip : The point of the toe farthest from the top of the landslide.

Tilt : 1. An observed change in the slope of the earth's surface. **2.** Angle of movement of fault blocks measured from normal.

Time domain : A seismic record is usually presented in the time domain, i.e., as a display of varying amplitudes of (filtered) ground motion as a function of time (in contrast to the equivalent representation in the frequency domain).

Toe - general : The break in slope at the foot of a bank where the bank meets the bed.

Toe - landslide : The lower, usually curved margin of the displaced material of a landslide. It is the most distant point from the main scarp.

Tongue : A long narrow strip of land, projecting into a body of water.

Topography : Surficial expression of earth's configuration including relief, position of natural and other features such as streams, roads, buildings, etc. Topography is indicative of features above sealevel (on continents) as well as below sealevel (in ocean basins).

Tornado : Extremely violent, localized storm with a characteristic funnel shape produced by extreme low pressure, enormous wind speeds and debris circulating around it. Wind velocities can reach 350km per hour. They form as funnel clouds from thunderstorms and hurricanes.

Torrential : Flow in a watercourse having a steep slope with great velocity and turbulence.

Transform fault : A variety of strike-slip fault that accommodates relative horizontal slip between other tectonic elements, such as oceanic crustal plates.

Tremor : Low amplitude, continuous earthquake activity.

Trench : 1. A narrow, elongate depression of the deep-sea floor, having steep sides and oriented parallel to the trend of an adjacent continent. It lies between the continental margin and the abyssal plain. Usually it forms the surficial trace of a subduction zone. **2.** A linear excavation made during mining or exploration activities.

Triple junction : Three fractures in the lithosphere, which meet at roughly 120° angles on a map. They form during the initial stages of continental rifting. Grabens usually form along the fracture patterns. As rifting continues, two of the three "arms" of the triple junction continue to widen, perhaps creating a new ocean basin between the separating continental pieces, while the third one "fails". The failed part is termed an aulocogen.

Tropical cyclones : Tropical cyclones are intense low-pressure systems, which form over warm ocean waters at low latitudes. Tropical cyclones are associated with strong winds, torrential rain and storm surges (in coastal areas). Tropical cyclones can cause extensive damage as a result of the strong wind, flooding (caused by either heavy rainfall or ocean storm surges) and landslides in mountainous areas as a result of heavy rainfall and saturated soil. They are also known as storms, gales, typhoons and hurricanes.

Trough : 1. A trough of low pressure is indicated by isobars extending outwards from an area of low pressure. The pressure at a point on a trough is lower than that at an adjacent point on either side of the line. **2.** In structural geology, portion of falling limbs of a fold is called trough. **3.** In oceanography, the lowest part of a wave, located between successive crests is also called trough.

Tsunami : Widely spaced, fast-moving ocean wave(s) initiated by sudden displacements of the sea floor during earthquakes or submarine landslides or volcanic eruptions.

Tsunami magnitude : A number used to compare sizes of tsunamis generated by different earthquakes; calculated from the logarithm of the maximum amplitude of the tsunami wave measured by a tide gauge distant from the tsunami source.

Tsunamigenic : Referring to those earthquake sources, commonly along major subduction-zone plate boundaries such as those bordering the Pacific Ocean that can generate tsunamis.

Tuff : A general term for consolidated rocks made of material ejected from volcanic explosions.

Tuff cone : A type of volcanic cone formed by the interaction of basaltic magma and water. It is smaller and steeper than a tuff ring.

Tuff ring : A wide, low-rimmed, well-bedded accumulation of hyaloclastic debris built around a volcanic vent located in a lake, coastal zone, marsh or area of abundant ground water.

Tumulus : A doming or small mound on the crest of a lava flow caused by pressure due to the difference in the rate of flow between the cooler crust and the more fluid lava below.

Turbidite : The sediments or rocks that formed as a result of a turbidity flow. Turbidity currents could get generated as a result of submarine earthquakes or slope failures or also by major rivers whose flow bypass continental shelf and supply sediments deep into slope and beyond.

Typhoon : Term used in the northwestern pacific for a tropical cyclone.

Upland : An area with higher relief relative to the surrounding areas, but not mountainous.

Uplift : The process or result of raising a portion of the Earth's crust through tectonic movement.

Valley : Depression in the land surface often occupied by a river.

Vei : The volcanic explosivity index, or vei, describes the relative size or magnitude of explosive volcanic eruptions. It is a 0-8 index of increasing explosivity. Each increase in number represents an increase around a factor of ten. The vei uses several factors to assign a number, including volume of erupted pyroclastic material (for example, ashfall, pyroclastic flows, and other ejecta), height of eruption column, duration in hours and qualitative/descriptive terms.

Vent : An opening in the earth's surface through which magma erupts or volcanic gases are emitted.

Ventifact : A stone that has been flattened and sharpened by wind abrasion. Ventifacts are commonly found strewn across a desert floor.

Vesicle : A small air pocket or cavity formed by escaping gas from lava during its solidification.

Volcanism : Describes the action or process of rising of magma and gases to the crust and being extruded onto the surface and into the atmosphere or intruded within rocks of lithosphere.

Volcanic activity : Outpouring of molten rock and other volcanic materials onto the Earth's surface. Outpouring may be either explosive or effusive.

Volcanic arc : *See* "Arc"

Volcanic Ash : Fine unconsolidated pyroclastic matter (less than 2mm in dia). It is called "tuff" when consolidated.

Volcanic breccia : Pyroclastic rock consisting of angular volcanic fragments that are larger than 64mm in diameter. The rock may or may not have matrix.

Volcaniclastic : A volcanic rock or unconsolidated deposit composed of pre-existing fragments, particles or clasts of volcanic origin.

Volcano : The solid structure created when lava, gases and hot particles escape to the Earth's surface through vents. Volcanoes are usually conical. A volcano is "active" when it is erupting or has erupted recently. Volcanoes that have not erupted recently but are considered likely to erupt in the future are said to be "dormant." A volcano that has not erupted for a long time and is not expected to erupt in the future is "extinct." These are all relative terms and have only limited validity. An extinct volcano can become active and similarly, an active volcano may go into dormancy for a while.

Vulnerability : The susceptibility of a geographic region for a geohazard due to its inherent characteristics such as proximity, geological or any other conditions, etc. to suffer harm.

Wave base : The bathymetry upto which waves may erode the bottom sediments. Fair weather wave base is the bathymetry upto which bottom sediments can be mobilized or eroded during normal weather conditions; Storm weather wave base is the bathymetry upto which the bottom sediments can be mobilized or eroded during storms and cyclones.

Wave climate : Average condition of the waves at a place, over a period of years, as shown by height, period, and direction, etc.

Wave direction : Direction from which a wave approaches.

Wave drift : The small net forward displacement of water in the direction of the wave travel, particularly in waves of large amplitude.

Wave front : The surface formed by a propagating wave (either seismic or water). The wave front is perpendicular to the seismic rays, which are oriented in the direction of wave propagation.

Wave height : The vertical distance between the crest (the highest point of a wave) and the trough (the lowest point).

Wave length : The distance between consecutive wave crests or wave troughs.

Wave period : Time in which a wave crest traverses a distance equal to one wavelength. It also indicates the time for two successive wave crests to pass a fixed point.

Wave propagation : The transmission of waves through a medium.

Wave refraction : The process by which wave direction is changed is called wave refraction. This process applies to oceanic and seismic waves. While the oceanic waves are refracted due to shoreline morphology and bottom topography, the seismic waves are influenced by density of the medium in which they travel.

Wave runup : The vertical distance above mean water level reached by the uprush of water from waves across a beach or up a structure.

Wave steepness : Ratio of wave height to wavelength.

Wave transformation : Change in wave energy due to the action of physical processes.

Wave-cut bench : A relatively level surface created by wave erosion along the coastline. A wave-cut bench stands above the water and extends seaward from the cliff.

Wave-cut platform : A horizontal bench of rock formed by wave erosion.

Weathering : The combination of physical, chemical and biological processes and agents by which exposure to atmospheric agents, such as air or moisture, causes rocks and minerals to break-down. This process takes place at or near the Earth's surface. Weathering entails little or no movement of the material that it loosens from the rocks and minerals.

Wind Erosion : Removal of soil particles by wind action.

Wind waves : The waves initially formed by the action of wind blowing over the sea surface. Wind waves are characterized by a range of heights, periods and wavelengths. As they leave the area of generation (fetch), wind waves develop a more ordered and uniform appearance and are referred to as swell or swell waves.

Zonation : A term used for delineation of a geographic region based on a specific character, namely, highly vulnerable zone or least vulnerable zone, etc.

References

Bates, R.L. and Julia, A.J., (Eds.), 1984 Dictionary of Geological Terms. 3rd Edition. American Geological Institute, Doubleday.

Shelton, D.C. and Prouty, D., 1979 Nature's Building Codes : Geology and Construction in Colorado. Colorado Geol.Surv.Spl.Publ. v.12.

Creath, W.B., 1996 Homebuyers' Guide to Geologic Hazards : An AIPG Issues and Answers Publication, Colorado Geological Survey.

Chatwin, S.C., Howes, D.E., Schwab, J.W. and Swanston, D.N., 1994 A Guide for management of landslide-prone terrain in the Pacific northwest, 2nd Edition. Crown Publications, Victoria, B.C.

Jochim, C.L., William, R.P., Truby, J.O., Wold, R.L. Jr., Weber, G. and Brown, S.P., 1988 Colorado Landslide Hazard Mitigation Plan, Colorado Geological Survey, Bulletin 48.

http ://mbmgquake.mtech.edu/seismic_glossary.html.

http ://www.napalladium.com/glossary.shtml.

http ://www.ucmp.berkeley.edu/glossary/glossary_0.html.

http ://www.igcworld.com/gm_glos.html.

Index

❑❑❑

FSC www.fsc.org
MIX
Papier | Fördert gute Waldnutzung
FSC® C083411

Zeitfracht Medien GmbH
Ferdinand-Jühlke-Straße 7
99095 Erfurt, Deutschland
produktsicherheit@kolibri360.de